Dictionary of Weighing Terms

Roland Nater · Arthur Reichmuth · Roman Schwartz
Michael Borys · Panagiotis Zervos

Dictionary of Weighing Terms

A Guide to the Terminology of Weighing

 Springer

Roland Nater
Mettler-Toledo International Inc.
8606 Greifensee
Switzerland

Arthur Reichmuth
Mettler-Toledo International Inc.
8606 Greifensee
Switzerland

Dr. Michael Borys
Physikalisch-Technische Bundesanstalt
Bundesallee 100
38116 Braunschweig, Germany

Dr.-Ing. Panagiotis Zervos
Physikalisch-Technische Bundesanstalt
Bundesallee 100
38116 Braunschweig, Germany

Dr. Roman Schwartz
Physikalisch-Technische Bundesanstalt
Bundesallee 100
38116 Braunschweig, Germany

ISBN 978-3-642-44425-8   ISBN 978-3-642-02014-8 (eBook)
DOI 10.1007/978-3-642-02014-8
Springer Dordrecht Heidelberg London New York

Pre-Press: Werner Brunner, Mettler-Toledo International Inc., Global MarCom Greifensee, Switzerland

Cover design: eStudio Calamar S.L.

Printed on acid-free paper

Springer is part of Springer Science+Business Media (www.springer.com)

Preface

This Dictionary of Weighing Terms is a comprehensive practical guide to the terminology of weighing for all users of weighing instruments in industry and science. It explains more than 1000 terms of weighing technology and related areas; numerous illustrations assist understanding. The Dictionary of Weighing Terms is a joint work of the German Federal Institute of Physics and Metrology (PTB) and METTLER TOLEDO, the weighing instruments manufacturer. Special thanks go to Peter Brandes, Michael Denzel, and Dr. Oliver Mack of PTB, and to Richard Davis of BIPM, who with their technical knowledge have contributed to the success of this work.

The Dictionary contains terms from the following fields: fundamentals of weighing, application and use of weighing instruments, international standards, legal requirements for weighing instruments, weighing accuracy. An index facilitates rapid location of the required term.

The authors welcome suggestions and corrections at www.mt.com/weighing-terms.

Braunschweig (DE) and Greifensee (CH),
The Authors   Summer 2009

Foreword

Since its founding in 1875, the International Bureau of Weights and Measures (BIPM) has had a unique role in mass metrology. The definition of the kilogram depends on an artefact conserved and used within our laboratories. The mass embodied in this artefact defines the kilogram, and this information is disseminated throughout the world to promote uniformity of measurements. Although the definition of the kilogram may change in the relatively near future, reflecting the success of new technologies and new requirements, the task of ensuring world-wide uniformity of mass measurements will remain.

But uniformity is not achieved through standards alone. In all areas of metrology, we seek a common language for referring to the apparatus we use, the rules we follow and the results we present. The field of mass metrology, or weighing, is vast and few of us have the time to become expert in all its areas. The Dictionary of Weighing Terms, with more than 1000 entries, will help bring clarity to this important area of metrology.

Dr R.S.Davis
Head, Mass Section   Summer 2009
BIPM

Contents

Index	1
Abbreviations	15
Encyclopedia	17
Literature References	263
Illustrations	269

Index

1999/92/EC 17
2003/94/EC 17
2004/10/EC 17
2004/108/EC 17
2004/22/EC 17
2004/9/EC 17
2006/42/EC 17
2006/95/EC 17
2009/23/EC 18
71/317/EEC 18
73/23/EEC 18
74/148/EEC 18
76/211/EEC 18
89/336/EEC 18
90/384/EEC 18
94/9/EC 19
98/37/EC 19
A/D converter 21
abbreviations 21
ability of being verified 21
Above-Medium Accuracy Weights Directive 21
absolute weighing 21
absorption 21
acceleration due to gravity 21
acceptable amount, smallest 22
accreditation 22
accuracy 22
accuracy class, higher 22
accuracy classes 22
accuracy classes of weighing instruments 23
accuracy classes of weight pieces 24
accuracy, medium 24
actual scale interval 24
adaptive filter 24
additive tare device 25
adjust, to 25
adjusting cavity 25
adjustment 25
adjustment weight 26
admission to verification 26
adsorption 26
AGME 26
air baggage scale 26
air buoyancy 26
air buoyancy correction 27

air damping 29
air density 29
air humidity 29
air pressure 29
alibi printer 31
ambient temperature 31
analog data processing device 31
analog error 31
analog output 31
analog readout 31
analog scale interval 32
analog signal 32
analog-digital converter 32
analytical balance 32
apparent mass 33
apparent weight 33
application 33
application module 33
application range of a weighing instrument 33
application temperature 33
apportion, to 33
Arbeitsgemeinschaft Mess- und Eichwesen 33
areometer 34
around-balance hanger 34
assembly 34
ASTM 34
ASTM International 34
ASTM weight classes 34
ATEX 35
ATEX 137 Directive 35
ATEX 95 Directive 35
auto-zero 35
autocal 35
automatic adjustment 36
automatic checkweigher 36
automatic conveyor 36
automatic gravimetric filling instrument 36
automatic inclination sensor 36
automatic instrument for continuous weighing 36
automatic instrument for discontinuous weighing 37
automatic rail scale 37
automatic release 37

automatic weighing instrument (AWI)	37
automatic zero maintenance	37
AutoMet	37
auxiliary device	38
auxiliary display	38
auxiliary indicating device	38
auxiliary indicator	38
auxiliary reading aids	38
auxiliary reading device	38
available capacity indicator	38
axis of action	38
axle-load scale	39
B	41
baby scale	41
back-weighing	41
balance	41
balance beam	41
balance for measuring surface tension	41
bar code	41
bar weight	41
base price	41
bathroom scale	41
beam balance	42
beam load cell	42
bearing	42
bed scale	42
below-the-balance weighing	42
belt loading	42
belt weigher	42
belt-conveyor scale	43
bench scale	43
Béranger scale	43
BEV	44
bias	44
bidirectional interface	44
BIML	44
bin scale	44
BIPM	44
Borda weighing method	45
Bouguer anomaly	45
bridge	45
bridge scale	45
bubble level	46
buoyancy	46
buoyancy force	46
burette	46
calibrate, to	49
calibration	49
calibration laboratory meeting ISO 17025	49
calibration service	49
calibration weight	49
canister load cell	49
carat scale	49
carat, metric	50
cash register systems	50
catch weigher	50
CE mark	50
CE marking for EC verification	50
CE year notation	51
center of gravity	51
certificate of conformity	51
certified computer	51
certified PC	51
characteristic curve	51
characteristic curve of a load cell	52
characteristic curve of a weighing instrument	52
checkout scale	52
checkweigher	52
CIPM	52
circular level indicator	52
classify according to mass, to	52
coarse dispensing	52
coarse display	52
coarse feed	53
coarse range	53
coarse weighing	53
coefficient of variation	53
combination scale	53
combined error	53
combined rail car and road vehicle scale	53
commercial scale	53
commercial weight	53
comparator balance	54
compensation coil	54
compensation current	54
compensation principle	54
compression column load cell	54
compression weighing cell	54
compulsory verification	54
computer, certified	54

confidence interval 54
confidence level 54
configuration 55
connecting hanger 55
connecting lever 55
constructional requirements 55
control chart 55
control limit 55
control unit 55
conventional mass 55
conventional scale interval 56
conventional value 56
conversion factor 56
converter 56
conveyor belt weigher 56
Coriolis mass counter 57
correction for air buoyancy 57
counter 57
counter scale 57
counterpoise weight 57
counting device 58
counting scale 58
coverage factor (k) 58
coverage interval 58
crane scale 58
creep error 59
cross-flexed bearing 59
cross-flexed spring joint 59
ct 59
current balance 59
customer keys 59
cylindrical weight 59
d 61
D/A converter 61
damping 61
damping device 61
damping systems 61
data bus 62
data concentrator 62
data matrix code 62
data memory 62
data plate 62
data storage device 63
data transmission 63
dead load 63
decimal balance 63
declaration of compatibility 64

declaration of conformity 64
deflection balance 64
deflection weighing device 64
degree of protection (IP) 64
degrees of protection provided by enclosures 64
Delta Range balance 65
DeltaRange (DR) 65
DeltaTrac 65
denier 65
denier balance 65
densitometer 65
density 66
density balance 66
density determination 66
density determination set 69
density of air 69
density of water 69
descriptive markings 69
design and function of a mechanical balance 69
design and function of an electro-dynamic balance 70
design and function of an electro-mechanical weighing instrument 71
Design Qualification 73
desorption 73
Deutscher Kalibrierdienst 73
deviation 73
dial weight 73
dial weight balance 73
dial weight combination 73
dialing step 73
dialysis scale 74
diet scale 74
differential eccentric load 74
differential linearity deviation 74
differential nonlinearity 74
differential weighing 74
digit 75
digital data processing device 75
digital device 75
digital display 75
digital filter 75
digital interval 75
digital printout 75
digital-analog converter 76

Directive on Above-Medium Accuracy Weights 76
Directive on Electromagnetic Compatibility 76
Directive on Machinery 76
Directive on Measuring Instruments 76
Directive on Medium Accuracy Weights 76
Directive on Non-Automatic Weighing Instruments 77
discrimination 77
dispenser 77
dispensing 77
dispensing balance 77
displacement body 77
display 77
display device 77
display device with reducible resolution 78
display error 78
display screen 78
division 78
division mark 78
DKD 78
draft shield 78
drift 78
drift of the measurement value 79
dry content 79
dryer 79
drying oven method 79
drying program 80
Dual Range 80
dual range balance 80
dual-range weighing instrument 80
dynamic axle-load scale 80
dynamic weighing 80
e 81
e-mark 81
EAN 81
EC Declaration of Conformity (DoC) 81
EC Directive 81
EC type approval 81
EC type examination 81
EC verification 82
EC verification mark 82
EC verification marking 82
eccentric load 82

eccentric load deviation 83
eccentric load test 83
eccentric load, differential 83
eccentric loading 83
eccentricity 83
eddy-current damping 83
EDP system 84
EDQM 84
effect 84
effective lever arm 84
effective mass 84
effective weight 84
electric charge 84
electrical safety 85
electrodynamic converter 85
electromagnetic compatibility (EMC) 85
electromagnetic force compensation 85
electromechanical weighing instrument 86
electronic assembly 86
electronic device 87
electronic weighing instrument 87
electrostatic charging 87
electrostatic discharging 87
electrostatic influence 87
EMC 87
EMC Directive 87
EMFC 87
EMFC load cell 88
EMFC weighing instrument 88
EMFR 88
EN 45501 88
EN 60529 88
endurance of the printout 88
engineering standards 88
environmental influence 88
equal-arm beam balance 89
equilibration of the weighing instrument 89
equilibrium 89
equilibrium position 89
Equipment Qualification 89
equivalence principle 90
error 90
error due to the display 90
error limit class 90
error limit component 90

error limits 90
error, random 91
error, systematic 91
European Declaration of Conformity 91
European Directive Concerning
Equipment and Protective Systems
Intended for Use in Potentially
Explosive Atmospheres 91
European Directive on
Electromagnetic Compatibility 91
European Directive on
Good Laboratory Practice 91
European Directive on
Good Manufacturing Practice 91
European Directive on
Measuring Instruments 91
European Directive on Non-Automatic
Weighing Instruments 91
European Directive on Prepackaged
Products 91
European Directive on requirements
for safety and health protection
of workers at risk from explosive
atmospheres 91
European Directive Relating to
Electrical Equipment Designed for
Use Within Certain Voltage Limits 91
European Directive relating to
medium accuracy weights 91
European Directive relating to
weights from 1 mg to 50 kg of
above-medium accuracy 92
European Machinery Directive 92
European Pharmacopeia 92
European Standard EN 45501 92
evaluation device 92
evaporation 92
exceptions to compulsory verification 93
expansion factor 93
explosion protection 93
extended displaying device 95
FACT 97
family 97
fill quantity 97
filling process control 97
filling process control facility (FPC) 97
filling scale 98

filter 98
filter balance 98
final weight value 99
fine adjuster 99
fine dispensing 99
fine feed 99
fine range 99
fine weight 100
firmware 100
flat-pan scale 100
flexible bearing 100
flexible coupling 100
flexible joint 100
flexure pivot 101
floor scale 101
fluid 101
foot switch 101
force 101
force comparison 101
force compensation 101
force due to gravity 101
force link 101
force measuring cell 102
forklift scale 102
form printer 102
formula weighing 102
formula weighing system 102
FPC 102
frame 102
G 103
gage factor 103
galvanic separation 103
gamma sphere 103
GAMP 103
garbage scale 103
gauge factor 103
Gaussian distribution 103
Gaussian weighing method 104
general approval 104
general clause 104
German calibration service 104
GLP 104
GMP 104
Good Automated Manufacturing
Practice 104
Good Laboratory Practice 105
Good Manufacturing Practice 105

gram	105
gravimetric	105
gravimetry	106
gravitation	106
gravitational attraction	106
gravity	106
gravity-dependent weighing instrument	109
Green M	109
gross value	109
gross weight	109
guide	109
guided pan	109
GxP	109
gyro load cell	110
gyro measurement cell	110
gyro scale	110
halogen lamp	111
hand scale	111
hanger	111
hanging load receptor	111
hanging pan	111
hardware	111
hierarchy of mass standards and weights	111
high-resolution	111
higher accuracy class	112
hopper scale	112
household scale	112
housing	112
hump scale	112
hybrid weighing instrument	112
hydrometer	113
hydrostatic balance	113
hygroscopic weighing sample	113
hysteresis	113
hysteresis compensation device	114
hysteresis deviation	114
identification mark	115
IEC	115
IEC 60529	115
ILAC	115
inclination	115
inclination error	115
inclination range	115
inclination sensor	115
inclination test	115
inclinometer	116
indication	116
industrial scale	116
influence of electrostatics	116
influence of environment	116
influence of humidity	116
influence of moisture	116
influence of temperature	116
influence quantities	116
infrared dryer	117
ingress protection	117
initial verification	117
initial zero-setting device	117
initial zero-setting range	117
inscriptions	117
inspection	118
installation of weighing instruments	118
Installation Qualification	119
integration time	119
integration time extension	119
interchange weighing method	119
interface	119
interference quantities	119
International Electrotechnical Commission	120
International Kilogram Prototype	120
International Laboratory Accreditation Cooperation	120
International Organization for Legal Metrology	120
International Organization for Standardization	120
International Prototype of the Kilogram	120
International System of Units	121
interpolation device	122
intervention limit	122
invariability	122
IP code	122
IP protection	122
ISO	122
ISO 17025	122
joint	123
joint flexure	123
k	125
kg	125
kilogram	125

kilogram prototype 125
knife-edge 125
knife-edge angle 125
knife-edge bearing 125
knife-edge plane 125
label printer 127
labels 127
laboratory balance 127
legal metrology 127
legal metrology requirements 127
legally relevant parameter 128
legally relevant software 128
letter scale 128
level 128
level indicator 128
level sensor 128
level, to 128
LEVEL-MATIC 128
leveling device 129
leveling screws 129
levelness compensation 129
lever 129
lever arm 130
lever arm, effective 130
lever chain 130
lever error 130
lever group 130
lever ratio 130
lever system 130
leverage 130
Lim 130
limit switch 130
limit value of inclination 131
limit value of tilt 131
limits of measurement errors 131
linearity 131
linearity deviation 131
linearization 131
liquid thermometer 131
LNE 131
load 131
load cell 132
load compensation 132
load drift 132
load lever 132
load limit 132
load pan 132

load range 132
load receptor 132
load relief device 133
load, eccentric 133
loading 133
local gravity 133
locking 133
locking device 133
long-term stability 134
long-term storage of measurement
data 134
low-resolution 134
Low Voltage Directive 134
low-level load receptor 134
low-level pan 134
low-profile scale 135
lumpiness of the weighing sample 135
LVD 135
machine 137
Machinery Directive 137
macroanalytical balance 137
magnetic damping 137
magnetic suspension balance 137
magnetism 138
magnetoelastic effect 138
main devices of the weighing
instrument 139
main verification mark 139
Maintenance Qualification 139
mass 139
mass attraction 139
mass comparator 139
mass comparison 140
mass counter 140
mass flow 140
mass normal 140
mass standard 140
mass, conventional 140
matrix code 140
Max 140
maximum capacity 140
maximum permissible deviation 141
maximum permissible error 141
maximum permissible error
in service 141
maximum permissible error
on verification 141

maximum permissible mass
difference 142
maximum rated load 142
maximum safe load 142
maximum tare 142
maximum tare effect 142
mean sample value 142
mean value 142
mean value trace 143
measurand 143
measurement bridge 143
measurement cylinder 143
measurement deviation 143
measurement mark 143
measurement pipette 143
measurement principle 143
measurement result 144
measurement result of a weighing 144
measurement signal 144
measurement time 144
measurement transducer 144
measurement uncertainty 144
measurement unit 145
measurement value 145
measurement value converter 145
measurement value deviation 145
measurement value drift 145
measuring chain 145
measuring container 145
measuring instrument 146
Measuring Instruments Directive 146
mechanical advantage 146
mechanical weighing instrument 146
medium accuracy 146
Medium Accuracy Weights Directive 146
METAS 146
method 147
method parameter 147
metric carat 147
metric system of units 147
metric ton 147
metric unit 147
metrological characteristics of
a weighing instrument 148
metrological test 148
metrological testing of weighing
instruments 148

metrologically relevant 148
metrology 149
metrology mark 149
mg 149
µg 149
microbalance 149
microdispenser 149
microgram 149
MID 149
milligram 149
Min 149
minimum capacity 149
minimum load 150
minimum sample weight 150
minimum weight 150
minus deviation 150
MinWeigh 150
modular concept 150
module 151
Mohr's balance 151
Mohr-Westphal balance 151
moisture content 151
momme 151
Monobloc 151
monolithic load cell 151
monorail scale 152
movable scale 152
mpe 152
Multi Range (MR) 152
multi-interval instrument 153
multi-pan balance 153
multi-range weighing instrument 153
multicomponent weighing
instrument 153
multihead weigher 153
multiple interval 153
multiple range 153
multiple range instrument 153
multiuser system 154
multiuser weighing system 154
n 155
N 155
National Conference on Weights
and Measures 155
National Institute of Standards
and Technology 155
national metrology institutes 155

National Type Evaluation Program 155
national verification mark 156
NAWI 156
NCWM 156
net value 156
net weight 156
newton 156
NIST 156
noise 157
nominal capacity 157
nominal fill quantity 157
nominal load 157
nominal load range 157
nominal range 157
nominal value 157
non-automatic weighing instrument 157
non-interacting data output 158
non-self-equilibrating instrument 158
non-self-indicating instrument 158
nonlinearity 158
nonlinearity, differential 158
nonmetric mass unit 158
nonmetric system of units 159
nonmetric unit 159
normal distribution 159
normal range 160
Notified Body 160
NPL 160
NTEP 160
number of scale intervals 160
number of verification scale intervals 161
numerical interval 161
obligation to record 163
Oechsle hydrometer 163
OIML 164
OIML certification system for measuring instruments 164
OIML recommendations and documents 164
OIML weighing instrument classes 164
OIML weight classes 164
onboard truck scale 166
operating modes of a weighing instrument 166
operating principle of a mechanical balance 166
operating principle of an electro-mechanical balance 166
operating principle of an electronic balance 166
operating temperature range of a weighing instrument 166
Operational Qualification 166
operator guidance 166
ordinal number 166
output signal 166
over/under scale 167
overhead rail scale 167
overload indicator 167
overload lock 168
overload protection 168
package 169
packaging 169
pallet scale 169
pan 169
pan brake 169
parallel guide 169
parallelogram 170
parcel scale 170
parts counting 170
passthrough sale 170
patient scale 170
pattern approval marks 170
pattern examination 171
PC, certified 172
percentage balance 172
Performance Qualification 172
period of verification validity 172
peripheral device 172
person scale 172
Pfanzeder scale 172
pharmacopeia 173
pharmacopoeia 173
physical weighing principle 173
pictogram 177
piece counting 177
piece counting system 177
piece-counting device 177
piezoelectric effect 177
piezoelectric scale 178
pin load cell 178
pipette 178
pivot joint 178

place of installation 178
place of use 179
place of verification 179
platform 179
platform scale 179
platter 179
PLU 179
plumb line 179
plummet 180
plunger 180
plus/minus balance 180
point of sale, public 180
pointer 180
poise beam 180
poise weight 180
position sensor 180
position vane 180
postal rate indicating machine 180
postal scale 181
power failure protection 181
PPD 181
precision 181
precision balance 181
precision weight 181
prepackage 181
prepackage process control 182
Prepackaged Products Decree (PPD) 182
Prepackaged Products Directive 182
prescription balance 182
preset tare device 183
pressure 183
preweighing 183
price indicator 183
price marker 183
price marker scale 183
price-computing weighing
instrument 183
primary display 183
print lock 184
printed record 184
printer 184
printer device 184
printing 184
printing device 184
printout 184
proFACT 184
program return after power failure 185

programmable or loadable software 185
projected scale 185
proportional weighing method 185
protected interface 185
protection type 185
prototype 185
PTB 186
public point of sale 186
public scale 186
purchase price 186
pycnometer 186
quality 187
quality assurance 187
quality control 187
quantity counting device 187
rail scale 189
rail wagon scale 189
random deviation 189
random error 189
range displacement 189
range switching 189
rapid drying procedure 189
rate indicating scale 190
ratio of mechanical advantage 190
readability 190
readiness 190
readout error 190
readout stabilization 190
receiving scale 190
reference current 190
reference density 191
reference mass 191
reference method 191
reference position 191
reference position of the weighing
instrument 191
reference voltage 191
reference weight 191
relative resolution 192
reliability 192
repairer identification mark 192
repairer identification stamp 192
repeatability 192
reproducibility 193
requirements for measuring
instruments 194
resolution 194

response threshold 194
rest position 194
rider 194
rider system 194
ring weight 195
road vehicle scale 195
Roberval scale 195
rocker pin 196
Roman beam scale 196
Roman dial scale 196
rope-tension scale 196
rounding error 196
rounding of measurement results 197
salesperson keys 199
sample 199
sample size 199
scale 199
scale cash register 199
scale division 200
scale interval 200
scale intervals, number of 200
scale mark 200
scale pit 200
scale spacing 200
scale value 200
scanning device 200
seal 200
seal, to 200
sealing 201
sealing point 201
securing sticker 201
sedimentation balance 201
self-adjustment 201
self-equilibrating instrument 201
self-indicating instrument 201
self-indication capacity 201
self-service weighing instrument 201
semi-self-equilibrating instrument 202
semi-self-indicating instrument 202
semimicro balance 202
sensitivity 202
sensitivity adjustment 203
sensitivity drift 203
sensitivity error 203
sensitivity offset 203
sensor 204
serial data transfer 204

set for the determination of density 204
set to zero 204
settling 204
settling position 205
settling time 205
shipping lock 205
SI units 205
signal 205
signal filter 205
signal processing 205
signal processing unit 205
significant 205
single component weighing instrument 206
single point load cell 206
Single Range (SR) 206
single range weighing instrument 206
single-pan balance 206
single-range balance 206
sinker 206
skip scale 206
sliding weight balance 207
slope 207
smallest acceptable amount 207
software 207
software identification 208
software securing 208
software separation 208
software, legally relevant 208
SOP 208
sort, to 208
sorting balance 208
specific weight 208
specification 209
spirit level 209
spring constant 209
spring element 209
spring force 210
spring measurement device 210
spring scale 210
SQC 210
stability 211
stability of the sensitivity 211
stability test 211
stabilization time 211
stamping label 211
stamping mark 211

stand-still	211
stand-still detector	211
stand-still lock	212
standard	212
standard deviation	212
standard gravity	212
standard load	212
standard measurement uncertainty	212
standard operating procedures	213
standard range	213
standard test package	213
standard test vehicle	213
standard uncertainty	213
standard weight	213
standard weight piece	213
standby operation	214
statistical confidence	214
statistical quality control	214
statistics	214
step method	215
strain gage	215
strain gage load cell	216
strain gage scale	216
strain gauge	216
strain gauge load cell	216
string	217
string balance	217
string load cell	217
subsequent verification	217
substitution balance	217
substitution weighing	218
subtractive tare device	218
suitability of a weighing instrument	218
support	218
surface tension	218
surface tension balance	219
switch-on behavior	219
switch-on drift	219
switchoff criterion	219
system scale	220
systematic deviation	220
systematic error	220
t	221
T	221
tael	221
tank scale	221
tare	221
tare compensation device	221
tare device	221
tare load	221
tare memory	222
tare signal	222
tare value	222
tare weighing device	222
tare weight	222
tare, to	222
target fill quantity	222
target value	222
taring material	222
taring range	223
taut band suspension	223
temperature compensation	223
temperature drift	223
temperature influence	223
temperature limits	224
temperature range	224
tendency correction device	224
tension weighing cell	224
tensitometer	224
terminal	224
test	225
test certificate	225
test load	225
test report	225
test weight	225
testing mark	226
tex	226
TGA	226
thermal analysis	226
thermal printer	226
thermobalance	226
thermogrammetry	226
thermogravimeter	227
thermogravimetric analysis (TGA)	227
thermogravimetry	227
three-knife balance	227
through-balance hanger	228
throughput	228
tilt	228
titration	228
tolerance	228
tolerance limit	228

top-loading 229
top-loading load receptor 229
torque balance 229
torsion balance 229
total control 230
traceability 230
tracing 230
triangular support 230
triboelectricity 230
triple-beam balance 230
truck scale 230
trueness 231
two-knife balance 231
type approval 231
type approval certificate 231
type evaluation 231
type examination 231
type label 231
type of protection 232
type-specific parameters 232
types of approval 232
ug 233
ultramicro balance 233
uncertainty (of a measurement) 233
uncertainty interval 233
underload indicator 233
unit 233
unit conversion factor 234
unit of force 234
unit of mass 234
unit of measurement 234
unit switching 234
unit symbol 234
United States Pharmacopeia 235
units law 235
unmodifiable software 235
UPC 235
USP 235
vacuum balance 237
validity period of verification 237
variability 237
variance 237
variation coefficient 237
vehicle on-board weighing system 237
vehicle scale 238
verifiable 238
verification 238

verification certificate 238
verification instructions 238
verification mark 238
Verification Ordinance 238
verification procedure for weighing instruments 239
verification scale interval 240
verification stamp 240
verification stickers 240
vibration 240
vibration damper 240
vibrations 240
vibrospatula 241
voltage fluctuation 241
voltage selector 241
volume 241
volume comparator 241
volume determination 242
volumetric 242
volumetric flask 242
volumetry 242
warm-on time 243
warm-up time 243
warning limit 243
water density 243
weigh in, to 244
weigh module 244
weigh out, to 244
weigh, to 244
weighbridge 245
weighed object 245
weighed-in quantity 245
weighed-out quantity 245
weigher 245
weighing 245
weighing boat 246
weighing capacity 246
weighing card 246
weighing chamber 246
weighing container 246
weighing deviations 246
weighing device 246
weighing error 246
weighing instrument 247
weighing instrument classes 248
weighing instrument construction 248
weighing instrument functions 248

weighing instrument of high
accuracy 248
weighing instrument of medium
accuracy 248
weighing instrument of ordinary
accuracy 248
weighing instrument of special
accuracy 249
Weighing Instruments Directive 249
weighing method 249
weighing pan 249
weighing piece 249
weighing rail insert 249
weighing range 249
weighing rate 249
weighing result 250
weighing room 250
weighing software 250
weighing speed 250
weighing system 250
weighing table 250
weighing terminal 250
weighing time 251
weighing tweezers 251
weighing uncertainty 251
weighing unit 251
weighing value 251
weighing-in aid 252
weighing-instrument-specific
parameters 252
weighing-out device 252
weighment 252
weight 252
weight class 253
weight classifier 253
weight effect 253
weight force 253
weight pan 253
weight piece 253
weight piece, cylindrical 254
weight pieces of higher accuracy
class 254
weight pieces of medium accuracy
class 254
weight set 254
weight unit 254
weight, specific 254

weight-dialing system 254
weightgrader for eggs 255
Weights and Measures Act 255
Weights and Measures approval 255
Weights and Measures authorities 255
Weights and Measures balance 255
Weights and Measures office 255
WELMEC 255
Westphal balance 255
Wheatstone bridge 255
wheel-load scale 256
wheel-load weigher 256
working standard 256
yarn balance 257
yarn count 257
year mark 257
year mark for national verification 257
year notation 257
year notation for national
verification 257
zero indicator device 259
zero load 259
zero mark 259
zero point 259
zero point correction device 259
zero point drift 259
zero point stability 259
zero position 259
zero-setting device 259
zero-setting range 260
zero-tracking device 260
zone of use 260

Abbreviations

The following abbreviations are used in this document:

→	cross-reference
a.k.a.	also known as
e.g.	for example
i.e.	that is
PC	Personal Computer
EC	European Community
EU	European Union
EEC	European Economic Community
EEA	European Economic Area

1999/92/EC

Directive 1999/92/EC of the European Parliament and of the
Council of 16 December 1999 on minimum requirements for
improving the safety and health protection of workers potentially at risk from explosive atmospheres. →ATEX 137 Directive

2003/94/EC

Commission Directive 2003/94/EC of 8 October 2003 lay-
ing down the principles and guidelines of →Good Manufacturing Practice in respect of medicinal products for human
use and investigational medicinal products for human use.

2004/10/EC

Directive 2004/10/EC of the European Parliament and of
the Council of 11 February 2004 on the harmonisation of
laws, regulations and administrative provisions relating to
the application of the principles of good laboratory practice
and the verification of their applications for tests on chemical
substances. →Good Laboratory Practice

2004/108/EC

Directive 2004/108/EC of the European Parliament and of
the Council of 15 December 2004 on the approximation of
the laws of the Member States relating to →electromagnetic
compatibility and repealing Directive →89/336/EEC.

2004/22/EC

Directive 2004/22/EC of the European Parliament and of
the Council of 31 March 2004 on measuring instruments.
→Measuring Instruments Directive

2004/9/EC

Directive 2004/9/EC of the European Parliament and of the
Council of 11 February 2004 on the inspection and verification of →Good Laboratory Practice (GLP).

2006/42/EC

Directive 2006/42/EC of the European Parliament and of
the Council of 17 May 2006 on machinery, and amending
Directive 95/16/EC.

2006/95/EC

Directive 2006/95/EC of the European Parliament and of
the Council of 12 December 2006 on the harmonisation of
the laws of Member States relating to electrical equipment
designed for use within certain voltage limits. →Low Voltage
Directive

2009/23/EC

Directive 2009/23/EC of the European Parliament and of the Council of 23 April 2009 on non-automatic weighing instruments. → Directive on Non-Automatic Weighing Instruments, → 90/384/EEC

71/317/EEC

Council Directive 71/317/EEC of 26 July 1971 on the approximation of the laws of the Member States relating to 5 to 50 kilogramme medium accuracy rectangular bar weights and 1 to 10 kilogramme medium accuracy cylindrical weights. → Directive on Medium Accuracy Weights

73/23/EEC

Council Directive 73/23/EEC of 19 February 1973 on the harmonization of the laws of Member States relating to electrical equipment designed for use within certain voltage limits. This directive was replaced by Directive → 2006/95/EC on 16 January 2007. → Low Voltage Directive

74/148/EEC

Council Directive 74/148/EEC of 4 March 1974 on the approximation of the laws of the Member States relating to weights of from 1 mg to 50 kg of above-medium accuracy. → Directive on Above-Medium Accuracy Weights

76/211/EEC

Council Directive 76/211/EEC of 20 January 1976 on the approximation of the laws of the Member States relating to the making-up by weight or by volume of certain prepackaged products. → Prepackaged Products Directive

89/336/EEC

Council Directive 89/336/EEC of 3 May 1989 on the approximation of the laws of the Member States relating to electromagnetic compatibility. This directive was replaced by Directive → 2004/108/EC on 20 July 2007.

90/384/EEC

Council Directive 90/384/EEC of 20 June 1990 on the harmonization of the laws of the Member States relating to → non-automatic weighing instruments. This directive was replaced by directive → 2009/23/EC on 5 June 2009. → Directive on Non-Automatic Weighing Instruments

94/9/EC

Directive 94/9/EC of the European Parliament and the Council of 23 March 1994 on the approximation of the laws of the Member States concerning equipment and protective systems intended for use in potentially explosive atmospheres. →ATEX 95 Directive

98/37/EC

Directive 98/37/EC of the European Parliament and of the Council of 22 June 1998 on the approximation of the laws of the Member States relating to machinery. This directive was replaced by directive →2006/42/EC on 29 December 2009. →Machinery Directive

A/D converter
→analog-digital converter

abbreviations
Weighing terms established by national and international regulations, standards, and agreements. Examples: →verification scale interval e, →number of scale intervals d, →maximum capacity Max, →minimum load Min, →accuracy class Ⓘ, Ⓘ, Ⓘ, Ⓘ or →number of verification scale intervals n.

ability of being verified
A measuring device or instrument (→balance, →weight piece) can be verified if it is generally approved for national verification or for →EC verification, if it satisfies the applicable verification requirements or if its design is approved for verification by the competent authorities. →admission to verification

Above-Medium Accuracy Weights Directive
Above-Medium Accuracy Directive →74/148/EEC

absolute weighing
Determination of the →mass or →conventional mass and indication of its measurement value in integrals, fractions, and multiples of the mass of the →International Prototype of the Kilogram. If greater accuracy is required when weighing in air, an →air buoyancy correction is necessary.

absorption
1. Process in which a solid body takes up another substance, a gas or a liquid, into itself. →weighing error (compare: →adsorption, →desorption)
2. Attenuation of electromagnetic radiation (radiation absorption) by transformation into heat. →physical weighing principle 3.1

acceleration due to gravity
If the surface that supports a body is removed, the body can fall freely. The →weight force that acts on the body causes it to accelerate. Since the inertial and gravitational →mass of a body are identical (→equivalence principle), the acceleration is equal to the →gravity and is given by

$$a = g \approx 9.81 \ \mathrm{m/s^2}.$$

The variation of →local gravity is primarily a function of the geographical latitude and elevation of the →place of installation.

acceptable amount, smallest
→smallest acceptable amount

accreditation
Formal recognition of the technical and organizational competence of a calibration, testing, inspection, or certification
laboratory to perform a specific service within the scope
of the accreditation according to internationally governing
standards. In many cases, accreditation is according to
ISO 17025 "General requirements for the competence of
testing and calibration laboratories".

accuracy
1. Closeness of agreement between a measured quantity value and a true quantity value of a measurand
   ([VIM:2008] 2.13).
2. Qualitative designation for the closeness of the approximation of determined results to the reference value.
   The reference value may be defined or agreed to be
   the true value or the expected value [DIN 55350-13].
   →error limits
3. The closeness of agreement between a test result and the
   accepted reference value ([ISO 5725] 3.6). Example:
   Ability of a measuring instrument to deliver output quantities that are close to the true value ([VIM:1993] 5.18).
   For repeated measurements, accuracy requires →true-
   ness (absence of →systematic errors) and →precision.
   For a single measurement, this need not necessarily be
   the case (Fig. 1).
4. The property of the stated values of weight pieces to correspond to their true value (→accuracy classes of weight
   pieces).
5. The property of the →measurement value of a weighing instrument to correspond to the value of the load on
   the instrument (→accuracy classes of weighing instruments).

accuracy class, higher
→higher accuracy class

accuracy classes
Classification of various types of →weighing instruments,
or →weight pieces, into classes of the same accuracy.
→weight classes, →accuracy classes of weighing instruments, →accuracy classes of weight pieces

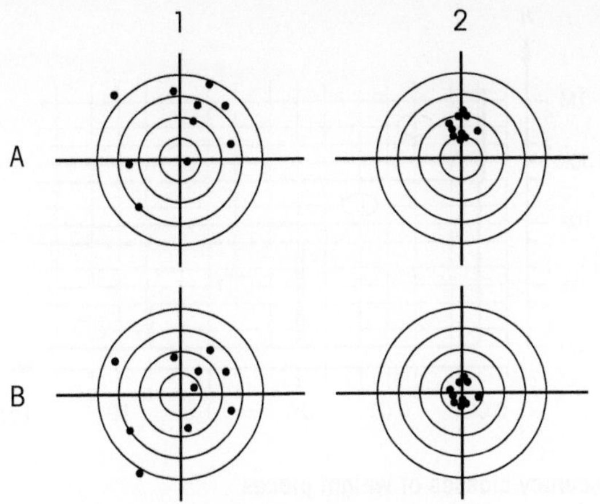

Fig. 1:
Explanation of the relationship between accuracy, precision, and trueness
Row A shows measuring points with systematic error (lack of trueness); row B measuring points with no systematic error (correct).
Column 1 shows scattered measurement points (lack of precision); column 2 shows measuring points with virtually no scatter (precise).
For repeated measurements, accuracy requires correct and precise measuring points; thus, in general, only the measuring points in field B2 are accurate.

accuracy classes of weighing instruments

Separation of →weighing instruments into different accuracy classes and assignment of different instrument types to classes of identical accuracy. The international recommendation [OIML R 76-1] and EC directives divide weighing instruments into the following accuracy classes (by decreasing accuracy): →Weighing instrument of special accuracy Ⓘ, →weighing instrument of high accuracy Ⓘ, →weighing instrument of medium accuracy Ⓘ, and →weighing instrument of ordinary accuracy Ⓘ ([OIML R 76-1] 3.1.1).
The →verification scale interval, →number of verification scale intervals and the →minimum capacity, in relation to the accuracy class of an instrument, are given in Tab. 1 and Fig. 2.

Tab. 1

Requirements for accuracy classes of weighing instruments according to OIML Recommendation R 76-1, 3.2. For instruments of class I with $d < 0.1$ mg, n may be less than 50000 (OIML R 76-1, 3.4.4).

Accuracy class	Verification scale interval e	Number of verification scale intervals $n = Max/e$		Minimum load [1]
		Minimum	Maximum	
I	$0.001\,g \le e$	50000	–	$100\,e$
II	$0.001\,g \le e \le 0.05\,g$ $0.1\,g \le e$	100 5000	100 000 100 000	$20\,e$ $50\,e$
III	$0.1\,g \le e \le 2\,g$ $5\,g \le e$	100 500	10 000 10 000	$20\,e$ $20\,e$
IIII	$5\,g \le e$	100	1000	$10\,e$

---

[1] With instruments of classes Ⓘ and Ⓘ, where the →scale interval d may be smaller than the →verification scale interval e, in the column "Minimum load" the verification scale interval e is replaced by the →scale interval d ([OIML R 76-1] 3.4.3).

Fig. 2
Number of verification scale intervals
n versus verification scale interval *e*
by accuracy class according to
OIML R 76-1

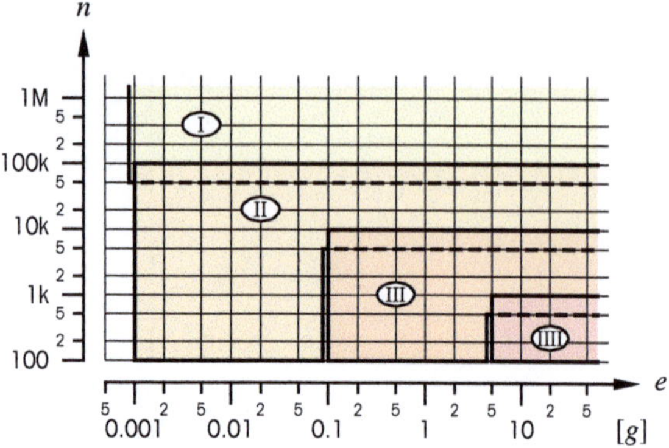

accuracy classes of weight pieces

Classes containing error limits for weights that are in systematic steps to fulfill different requirements. In most cases the error limits proceed from class to class in logarithmic steps, i.e. for each step the relative error limit increases (or decreases) by the same factor. Examples of the assignment of weight pieces to →accuracy classes are:

1. →OIML weight classes (international)
2. →ASTM weight classes (USA)
3. Medium accuracy weights (→Directive on Medium Accuracy Weights, →71/317/EEC)
4. Weights of above-medium accuracy (→Directive on Above-Medium Accuracy Weights, →74/148/EEC)

accuracy, medium

→medium accuracy

actual scale interval

Value expressed in →units of mass

1. for analog indication or analog printout: difference between the values corresponding to two consecutive scale marks (→scale division); scale interval d.
2. for digital indication or digital printout: difference between two consecutive indicated values (→numerical interval); digital scale interval d.
   ([OIML R 76-1] T.3.2.2)
3. A fixed value (→conventional scale interval d) used to classify weighing instruments that are not equipped with an indicator device.
   →readability

adaptive filter

→Signal filter with variable characteristics, usually implemented as a →digital filter. Adaptive filters modify their

characteristics automatically according to the weighing
signal to attain an optimum between interference suppression and →settling time.

additive tare device
A device used to weigh or compensate a →tare load without utilizing any part of the weighing range of the balance.
→tare weighing device, →tare compensation device
(compare: →subtractive tare device)

adjust, to
1. Set of operations carried out on a measuring system so
   that it provides prescribed indications corresponding to
   given values of a quantity to be measured ([VIM:2008]
   3.11).
2. Adjusting is the action of setting a measuring instrument
   or standard so that the measured value is correct, or
   deviates as little as possible from the correct value, or the
   deviation remains within acceptable limits of error.
   This is obtained
   a) in the case of a weighing instrument, through ad-
      justing the manual fine setting of its →indication by
      trained specialist personnel, or semi-automatically
      by the user, by placing on the instrument a →refer-
      ence weight that is kept either externally or inside the
      instrument, or automatically if the instrument has an
      adjusting mechanism with reference weight (→self-
      adjustment).
   b) in the case of a weight piece, through correcting its
      mass to the corresponding nominal value, e.g. by
      filing, or by adding or removing correction material in
      an →adjusting cavity.
   (compare: →calibration)

adjusting cavity
Sealable cavity, a.k.a. adjustment cavity.
1. In weight pieces to hold the material used to adjust
   the weights to their nominal value (Fig. 182b). Weight
   pieces of OIML →accuracy class E1 and E2, and ASTM
   class 0 (→accuracy classes of weight pieces) are not
   permitted to have adjusting cavities.
2. On the load receptor of a mechanical weighing instrument to set the unloaded instrument to read zero.

adjustment
1. Result of the action of adjusting (→to adjust) an instrument.

2. Non-technical simplified term for →sensitivity adjustment.

adjustment weight
→reference weight

admission to verification
Requirements specified in the →Weights and Measures Acts
for the verification of measuring instruments. A type may
either be generally approved, or approved after testing by the
→Notified Body. →types of approval

adsorption
Process in which a liquid or gaseous substance is retained
on the surface of a solid body. →weighing error (compare:
→desorption, →absorption)

AGME
Abbreviation for →'Arbeitsgemeinschaft Mess- und Eichwesen'.

air baggage scale
→Scale to determine the weight of passengers' baggage
with readout possibilities on two sides and a load surface
that allows easy handover of the baggage (Fig. 3).

air buoyancy
Buoyancy force that counteracts the →weight force of a body
that is surrounded by air. The magnitude of the →buoyancy
force is given by

$$F_a = m_a g = \rho_a V g = \frac{\rho_a}{\rho} mg$$

where
- m mass of the body
- ρ density of the body
- ρ_a density of the air (→air density)
- V volume of the body
- m_a mass of the air displaced by the body
- g →local gravity.

Since the buoyancy force of a body that is involved in a
weighment cannot be separately detected by the weighing
instrument, the instrument does not indicate the value of
the →mass, but the →weighing value. Air buoyancy is
usually the main cause of →systematic error when weighing
in air, particularly in →high-resolution weighing.
→deviation

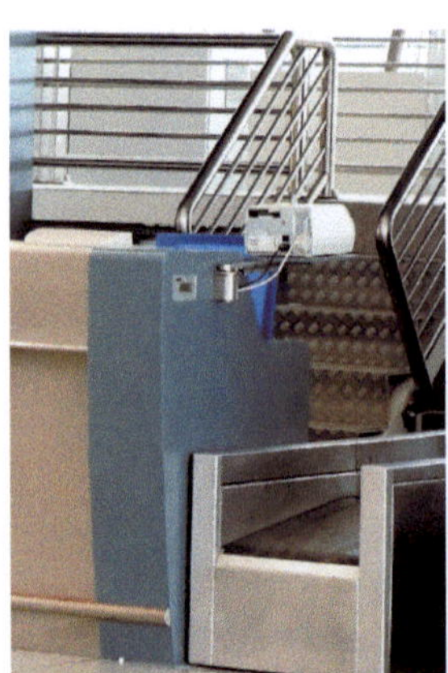

Fig. 3
Air baggage scale

air buoyancy correction

A →weighment performed in air is subject to →air buoyancy. Normally, the →weighing value does not contain
a correction for air buoyancy. Unless stated otherwise,
→weighing instruments and →weight pieces are always
adjusted to the conventionally defined reference density of
$\rho_c = 8000$ kg/m^3 (→conventional mass).

1. If the purpose of the weighment is to determine the
   →mass m of the weighed object, the →weighing value
   W (the sum of the weight pieces, the value read from, or
   indicated by, the weighing instrument) must be multiplied
   by the factor B_W, i.e.

$$m = B_W\, W$$

where

$$B_W = \frac{1 - \dfrac{\rho_a}{\rho_c}}{1 - \dfrac{\rho_a}{\rho}}$$

In this formula are

ρ_a density of the air (at the time of weighing)
        (→air density)

ρ_c conventional object density 8000 kg/m^3

ρ density of the weighed object.

The correction for air buoyancy can be obtained from
Fig. 4.

2. If the purpose of the weighment is to determine the
   →conventional mass m_c of the weighed object, the
   weighing value W (the sum of the weight pieces, the
   value read from, or indicated by, the weighing instrument) must be corrected as follows:

$$m_c = \frac{1 - \dfrac{\rho_a^*}{\rho_r}}{1 - \dfrac{\rho_a}{\rho}} \; \frac{1 - \dfrac{(\rho_a)_c}{\rho}}{1 - \dfrac{(\rho_a)_c}{\rho_r}} \; W$$

In this formula

ρ_a density of the air at the time of weighing
        (→air density)

ρ_a^* 1. for weighing instruments that function by mass
        comparison (→physical weighing principle 1):
        density of the air at the time of weighing: $\rho_a^* = \rho_a$
        2. for weighing instruments that function by force
        comparison (→physical weighing principle 2):
        density of the air at the time of the →sensitivity
        adjustment: $\rho_a^* = (\rho_a)_r$

ρ_r density of the reference weights
        (if not known, use $\rho_c = 8000$ kg/m^3)

$(\rho_a)_c$ conventional density of air 1.2 kg/m^3

ρ density of the weighed object.

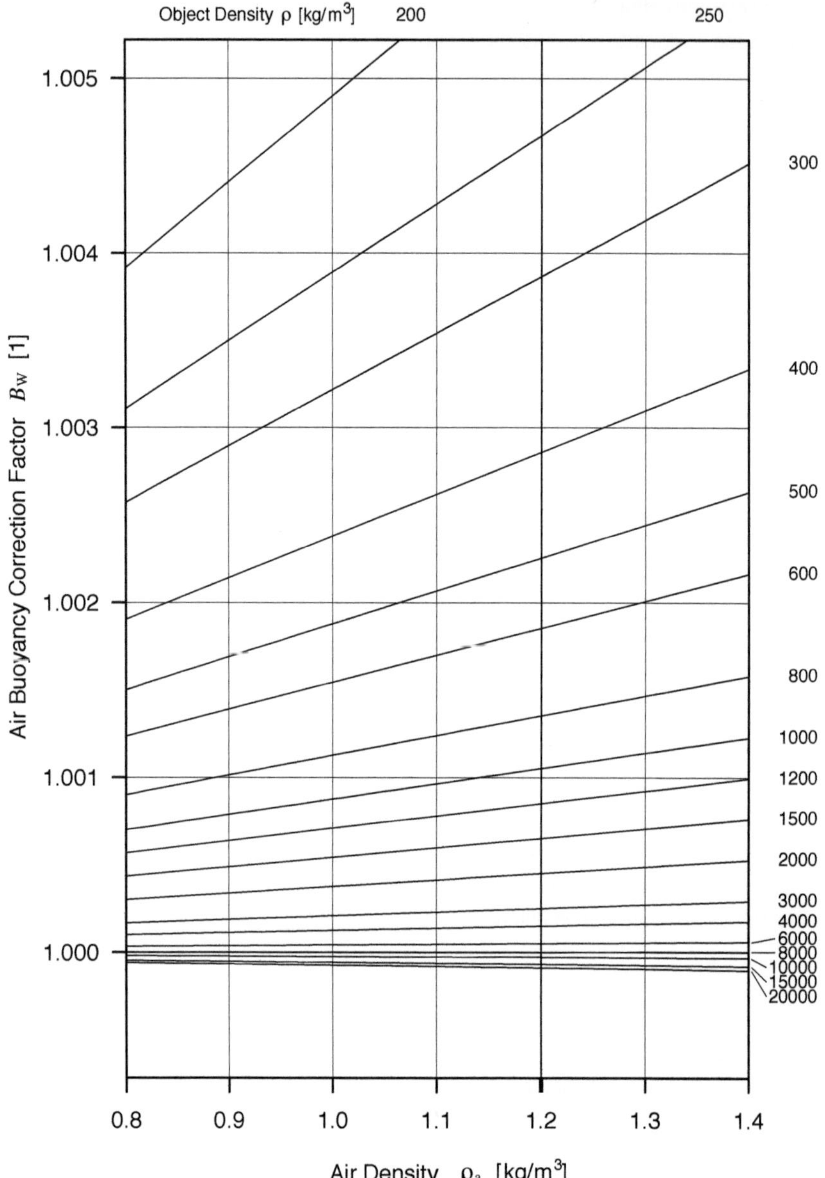

Fig. 4
Correction for air buoyancy: To obtain the mass of the weighed object, the value read from the weighing instrument must be multiplied by the appropriate factor.

$ρ_a$: air density
$ρ$: density of the weighed object
B_W: air density correction factor

air damping

→damping systems

air density

The →density ρ_a of dry air is proportional to the →air pressure and inversely proportional to the absolute temperature, according to

$$\rho_a = \frac{1}{R_a}\frac{p}{T}$$

R_a gas constant of dry air: ~ 287 J/(kg·K)
p air pressure [2]
T absolute air temperature [K].

Under normal conditions at sea level (20 °C, 1013 hPa), the density of dry air is therefore

$$(\rho_a)_0 = \frac{1}{287\,\frac{J}{kg\cdot K}}\,\frac{1013\ \text{hPa}}{(20 + 273.15)\text{K}} = 1.20\,\frac{\text{kg}}{\text{m}^3}$$

If the density of air needs to be determined more accurately, its humidity must also be taken into account

$$\rho_a = \frac{0.348444\,p - (0.00252\,t - 0.020582)\,h}{273.15 + t} \quad ^{[3]} \tag{1}$$

ρ_a air density [kg/m³]
p air pressure [2] [hPa]
h relative air humidity [%]
t air temperature [°C].

In the range of (1.2 kg/m³) ±10%, the air density determined with formula (1) has a typical relative uncertainty of 4×10^{-4}.

air humidity

The amount of water vapor in the air. Relative humidity h is the ratio between the actual vapor pressure of water and its saturated vapor pressure.

air pressure

The static →pressure prevailing in the mixture of gases that forms the Earth's atmosphere. Mean air pressure at sea level is 1013 hPa (normal pressure) and decreases continuously with increasing height. It also fluctuates constantly with changing weather conditions. The standard deviation of these fluctuations from the local mean value over a period of two or more weeks at temperate latitudes is of the order of 7 hPa, or approximately 0.7% of normal pressure.

---

[2] Station pressure STP (→air pressure)
[3] Simplified version of CIPM-formula, standard version ([CG-18], Appendix A, Formula A1.2-1)

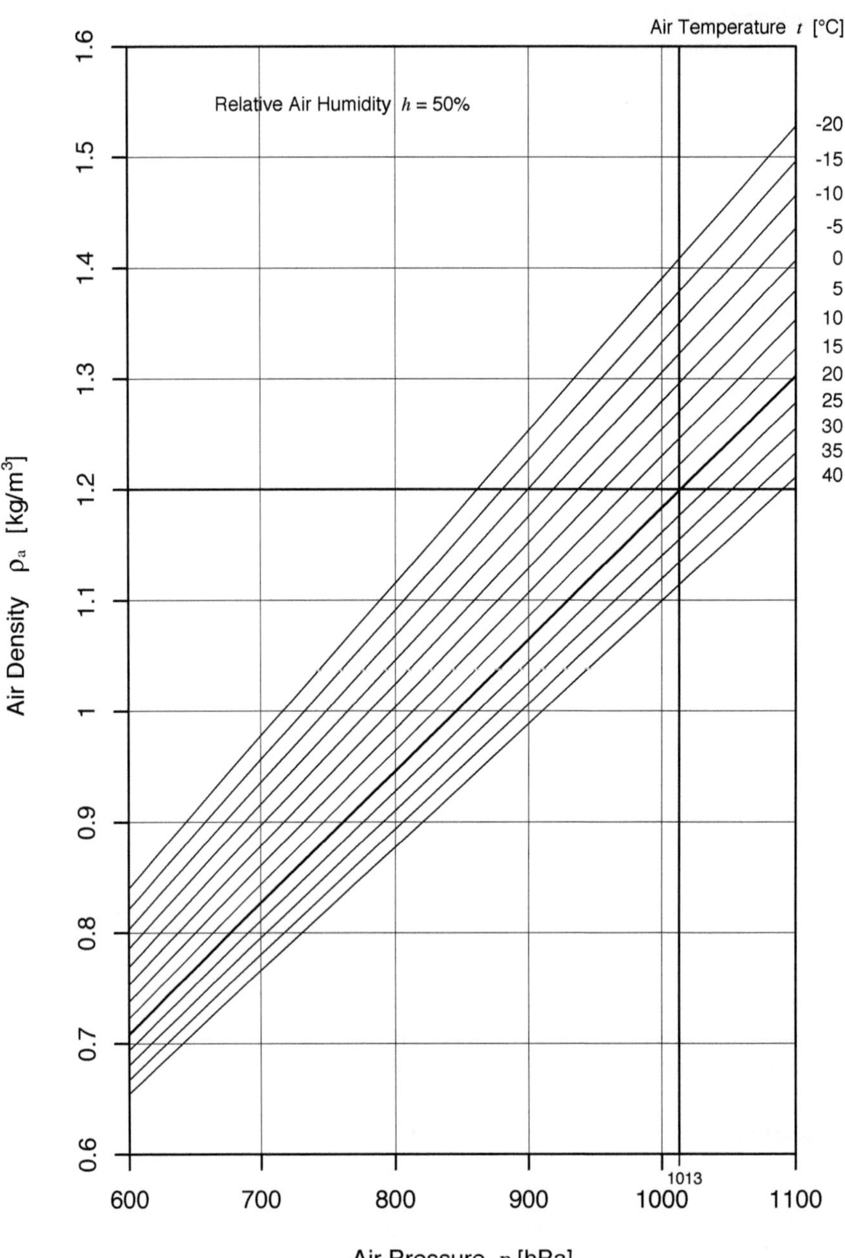

Fig. 5
Air density as a function of air pressure and temperature [according to formula (1)]

p: air pressure
ρ_a: air density
t: air temperature
$h = 50\%$: relative air humidity

In meteorology, the local air pressure is referred to as "station pressure", which is abbreviated to STP or QFE. On weather charts, however, the air pressure reduced to sea level is shown, which is indicated by the abbreviation SLP (sea level pressure) or QFF. If the density of air is to be determined for the correction for air buoyancy (→air buoyancy correction), station pressure (STP or QFE) must be used.

alibi printer

Non-technical expression for a →verifiable printing device in a verified →weighing system to which additional non-verified auxiliary devices or data processing systems with printer are connected.

ambient temperature

Temperature of the air surrounding the →weighing instrument and weighed object. →temperature range, →temperature drift

analog data processing device

→Electronic device in a →weighing instrument that performs the analog-digital conversion (→analog-digital converter) of the →output signal of the →load cell, performs further processing of these data, and forwards the weighing results in digital form across an interface, but without indicating them. ([OIML R 76-1] T.2.2.3) (compare: →digital data processing device)

analog error

Error due to the analog display.

analog output

An electrical output from a weighing instrument where, for instance, the →measurement value can be represented by an electrical voltage or current whose value changes continuously as the load changes. This output allows the connection of electrical measuring instruments and recorders which indicate results in analog manner. →analog signal.

analog readout

→Measurement values are continuously indicated by the position of an index mark (line, →pointer) against a line scale which is generally marked with numbers (→scale) (Fig 6). The analog readout makes it possible to determine measurement values in fractions of the division value. This can, however, give rise to subjective reading errors.

Fig. 6
Example of an analog readout

Fig. 7
Analog-digital converter
(left: analog input signal,
right: digital output signal)

analog scale interval
→actual scale interval

analog signal
A stepless variable, usually electrical quantity (e.g. current, voltage) that is proportional to the →measurement value. →analog output

analog-digital converter
An →electronic device for converting →analog signals (voltages, currents) into digital signals, e.g. in a digital voltmeter (Fig. 7). Often referred to as A/D converter. (compare: →digital-analog converter)

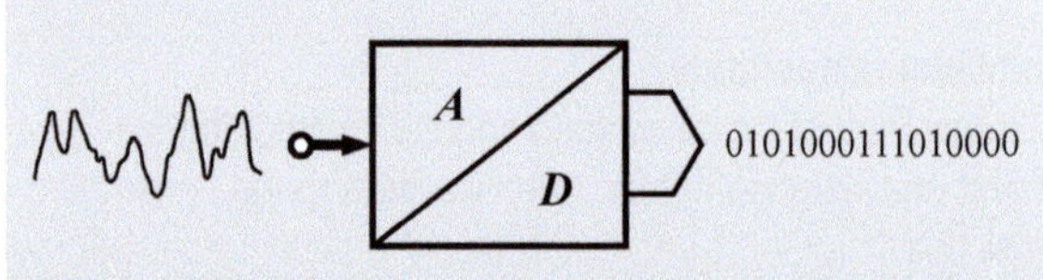

analytical balance
1. Collective term for →weighing instrument of special accuracy with high →resolution of accuracy class ① (→accuracy classes of weighing instruments). Analytical →balances are subdivided into →macroanalytical balances, →semimicro balances, →microbalances, and →ultramicro balances. Although the →weighing capacity and →readability of these balances were originally selected to be suitable for chemical analyses, such balances are used wherever there is a requirement for high →resolution and →accuracy. On an analytical balance a small quantity can be weighed in a relatively heavy container, which requires a readability of 0.1 mg or less and at least 100000 →actual scale intervals (→number of scale intervals). Because of their high →resolution (i.e. small readability), analytical balances are fitted with a →draft shield to protect them from disturbing air currents.
2. Strictly by definition: →Weighing instrument of accuracy class ① (→accuracy classes of weighing instruments) with a →weighing capacity in the range of 100…400 g (typically 200 g) and a →readability of 0.1 mg. Also referred to as →macroanalytical balance (Fig. 8).
3. Strictly by definition: Weighing instrument according to 2. that satisfies the corresponding →legal metrology requirements.

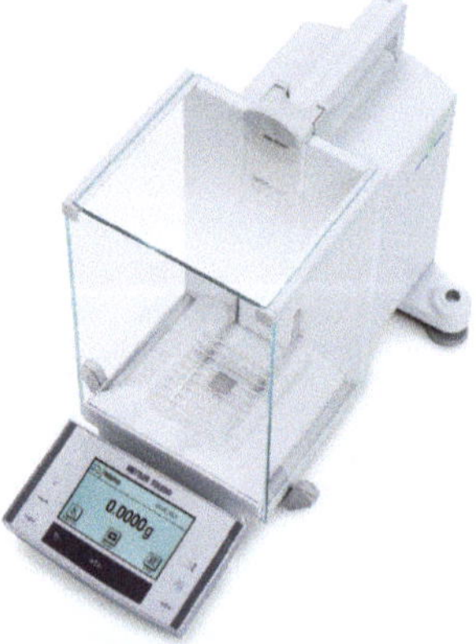

Fig. 8
Analytical balance (weighing capacity 200 g; readability 0.1 mg)

apparent mass

Ambiguous term sometimes used for →effective mass or for →weighing value.

apparent weight

Ambiguous term sometimes used for →effective weight.

application

1. Weighing procedure, weighing method. →weighing instrument functions
2. →application module, →weighing software

application module

A program module that may be either external or built into the weighing instrument, and may or may not have an additional program memory or additional keyboard, to control certain predefined weighing procedures or applications. Examples: percentage weighing (→percentage balance), →piece counting, →formula weighing, →statistics (Fig. 9), →density determination, →dynamic weighing.

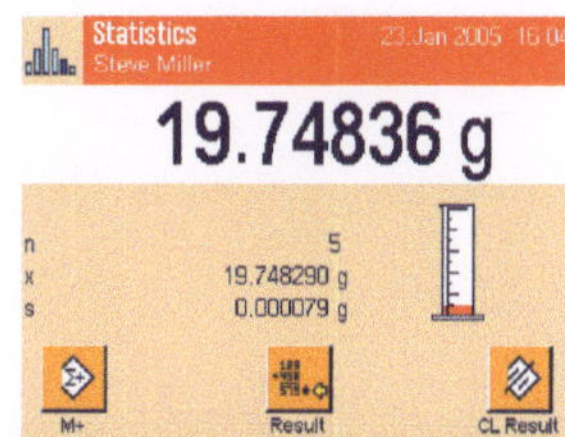

Fig. 9
Application module for statistical applications

application range of a weighing instrument

Restrictions regarding the intended use and/or environmental conditions under which a →weighing instrument may be used. Examples: "Scale/balance not to be used in public points of sale", or special →temperature limits "–10°C to 40°C".

application temperature

The temperature set on a →dryer when defining a →method. Because of the varying radiation →absorption of the sample, the temperature of the sample may differ slightly from the set temperature.

apportion, to

Using a weighing instrument to separate designated quantities of merchandise according to mass (as opposed to →weighing). If apportioning is automatic, the →weighed object is automatically conveyed to the →load receptor and automatically separated into equal quantities, e.g. to produce →prepackages.

Arbeitsgemeinschaft Mess- und Eichwesen

Coordinating body of the Weights and Measures authorities of Germany's 16 federal states (www.agme.de).

areometer
Synonym for →'hydrometer'.

around-balance hanger
A device to which other →load receptors can be attached below the weighing chamber or bottom of the balance for →below-the-balance weighing (Fig. 10).

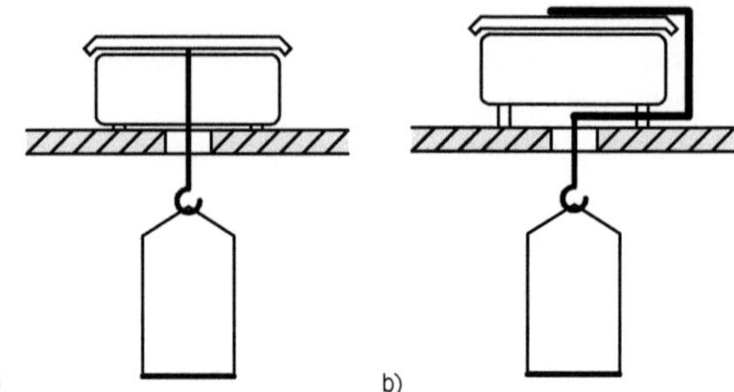

Fig. 10
Below-the-balance weighing
a) through-the-balance hanger;
b) around-the-balance hanger

assembly
→module

ASTM
→ASTM International

ASTM International
An organization in the United States of America that develops standards as well as related technical information and services that are globally recognized. The organization was formerly known as the American Society for Testing and Materials (ASTM).

ASTM weight classes
Separation of weights into classes according to error limits that are specified in ANSI/ASTM guidelines. ANSI/ASTM E617 [ASTM E 617] "Specification for Laboratory Weights and Precision Mass Standards" defines the characteristics for weights from 1 mg to 5 t in eight classes 0, 1, 2 to 7; →OIML Recommendation R 111 (→weight classes) is also recognized. The maximum relative permissible error (→maximum permissible error, mpe) for weights of class 0 is $1.3 \cdot 10^{-6}$ (for weights $\geq$ 100 g) and reduces per three classes by approximately a factor of 10 to approximately 0.05% (1 kg) for class 7. The weights may be of any shape provided that it does not interfere with their reliability. On the other hand, the materials to be used and their densities, the surface qualities, the magnetic characteristics, etc. are specified for each class. The calibration uncertainty U must

not exceed $^1/_3$ of the *mpe* at $k = 2$, which corresponds to a
→standard uncertainty u of $^1/_6$ *mpe*. The deviation of the
→mass or →conventional mass from the nominal value
must not occupy more than the remainder of the *mpe*. The
nominal values in →SI units are regarded as preferences.

ATEX

Abbreviation for the French term 'atmosphère explosible'
(explosive atmosphere). →94/9/EC, →1999/92/EC,
→explosion protection

ATEX 137 Directive

This European Directive regulates the organizational and
technical measures for the operators of plants to ensure that
in hazardous areas there is no safety risk from explosions
for persons working in those areas (→explosion protection).
The operator of a plant must prepare a risk analysis of
potential explosion hazards and specify measures to reduce
the hazards to an acceptable level (explosion protection
document). Equipment, components, and protective systems
intended for use in hazardous areas must be installed, operated, and maintained appropriately and according to the
manufacturer's instructions. The directive is implemented
as national law in the EEA. In Switzerland, the directive is
essentially adopted in an information sheet of the Swiss
Accident Insurance Fund (SUVA 2153). Each country is
entitled to define further measures. →1999/92/EC,
→ATEX 95 Directive

ATEX 95 Directive

This European Directive regulates the measures to ensure
that equipment, components, and protective systems that
are intended for use in hazardous areas may only be put
into operation and installed provided that, with appropriate
installation and maintenance, they do not present a hazard
to people (→explosion protection). In addition, the manufacturer must ensure that the affected installations comply
with the so-called "essential health and safety requirements"
that are listed in the directive. The directive is implemented
as national law in the EEA and Switzerland. →94/9/EC,
→ATEX 137 Directive

auto-zero

→zero-tracking device

autocal

Collective term for fully automatic monitoring and →sensitivity adjustment. On occurrence of →sensitivity drift caused by change of location, →drift (over time), →temperature drift, etc., a recalibration is performed automatically according to plan (time schedule) or automatically. →automatic adjustment, →FACT, →proFACT

automatic adjustment
Device for automatic →adjustment of the →sensitivity of →weighing instruments. The adjustment operation can be performed, for instance, by pressing a button, or automatically after a certain period of time, or by a change of temperature. Initiation of the adjustment causes motorized placement of a built-in reference mass on the weighing instrument. →FACT, →proFACT, →self-adjustment

automatic checkweigher
An →automatic weighing instrument that makes it possible to determine whether a package filled with the same →nominal fill quantity lies within or outside preselected limits (Fig. 11). An upstream filling machine can be adjusted by means of an attached tendency correction device or a control program.

automatic conveyor
Device that is mainly used to automatically feed items such as capsules and tablets onto a balance or weighing system.

automatic gravimetric filling instrument
→Automatic weighing instrument used to obtain equal, preselected fill quantities. Separated into coarse and fine feeds, the material to be weighed is conveyed to the weighing instrument by means of special transporting devices. Depending on the type of →load receptor, automatic weighing instruments can be purchased with an emptying device (e.g. hopper scales) or without such a device (bag-filling scales).

automatic inclination sensor
Device that measures the deviation of the →axis of action of a weighing instrument from the vertical (→inclination) and triggers an alarm signal or displays a corresponding message when a limit value is violated. Alternatively, the information about the inclination can also be used to correct its effect on the weighing signal. ([OIML R 76-1] 3.9.1.1, →limit value of inclination

automatic instrument for continuous weighing
Automatic weighing instrument for →weighing an uninter-

Fig. 11
Automatic checkweigher

rupted flow of material without it being systematically subdivided (e.g. →conveyor belt weigher).

automatic instrument for discontinuous weighing
An automatic instrument used →to weigh materials of different mass, in some cases by summing several individual
weighing results.

automatic rail scale
→Automatic weighing instrument that has a →load recep-
tor with rails on which rail vehicles can travel ([DIN 8129]
T.1.3) (Fig. 12). →hump scale

automatic release
Contrary to normal manual release (→locking device) on
→mechanical weighing instruments, with automatic release
the process takes place according to a fixed time program
regardless of the actuation speed of the →shipping lock.
Automatic release results in an improvement in reproducibility and protects the weighing-out device and its assemblies
against shocks.

automatic weighing instrument (AWI)
A →weighing instrument that performs weighing procedures
without the intervention of an operator and continuously
reinitiates automatic weighing procedures that are characteristic of the instrument. The following are types of automatic
weighing instruments:
→automatic gravimetric filling instrument
→automatic instrument for discontinuous weighing
→automatic instrument for continuous weighing
   (→belt weigher)
→automatic checkweigher
→automatic rail scale
→weightgrader for eggs
(compare: →non-automatic weighing instrument)

automatic zero maintenance
→zero-tracking device

AutoMet
A test measurement with a →dryer which at a selected
temperature, and with a selected drying program and sample quantity, determines the →switchoff criterion at which
the measurement value most closely matches the reference
value.

Fig. 12
Rail scale with a weighing capacity of
up to 400 t

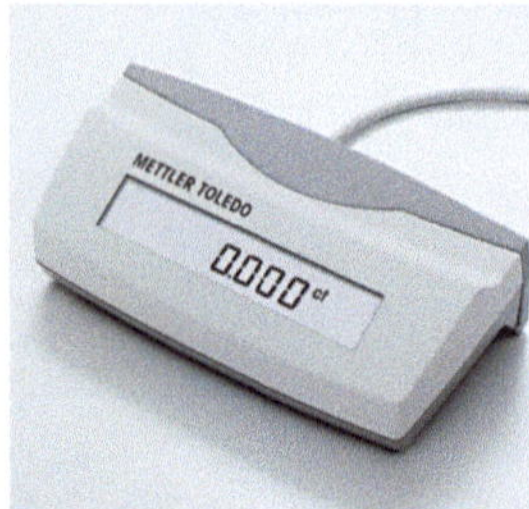

Fig. 13
Auxiliary display

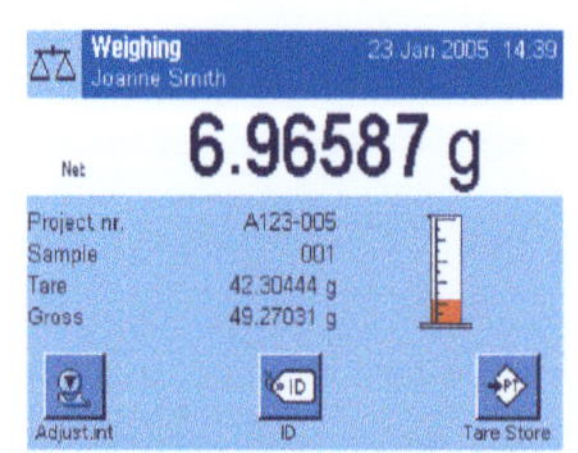

Fig. 14
Available capacity indicator: The
measuring beaker is approximately
one-fifth full, which indicates that
approximately 40 g are on the
balance (which has a maximum
capacity of 200 g).

auxiliary device

A device that is connected to, or mounted on, the weighing
instrument to provide additional presentation, forwarding, or
processing of the →weighing results and other →primary
displays, e.g. →printer, →digital display, →terminal, →data
memory, PC.

auxiliary display

A display that is additional to the →primary display
(Fig. 13). →auxiliary device

auxiliary indicating device

Device for displaying figures on verifiable weighing instruments of →accuracy classes ① and ⑪ whose lowest-
value figure is clearly differentiated from the other figures. In
the specially designated display position, no →verification
scale intervals e, only →actual scale intervals d, may be
displayed. A →multi-interval instrument must not be fitted
with an auxiliary indicating device. ([OIML R 76-1] 3.4)

auxiliary indicator

In verified operation, these are displays, signals, and symbols that are not →primary displays.

auxiliary reading aids

Allow the →weighing result to be read with higher →resolu-
tion. Collective term for →fine adjuster, →interpolation device, →auxiliary indicating device.

auxiliary reading device

Device on division →scales to reduce the reading uncertainty (e.g. vernier, →fine adjuster) or additional display
position whose →division (d) is less than the →verification
scale interval (e). →display device with reducible resolution,
→interpolation device

available capacity indicator

An additional →display device on a weighing instrument
for the rapid approximate determination of the →weighing
result or for observation of a loading operation (Fig. 14).
→dispensing

axis of action

The intended axis of use of a →load cell or →weighing
instrument on which the weight force to be measured should
lie and on which the →sensitivity of the load cell is at its
maximum (→reference position). The axis of action of a

weighing instrument is perpendicular to the →load receptor and usually passes through its center; the axis of action of a load cell is usually identical, or parallel, to a main axis. Forces that are applied outside the axis of action cause →eccentric load deviation, forces that are not parallel to it cause →inclination error. →effective lever arm

axle-load scale
A scale for measuring the load on the individual axles of a road vehicle (Fig. 15).

Fig. 15
Axle-load scale
(Image by courtesy of
Gassner Wiege- und Messtechnik,
Salzburg, AT)

B
Symbol for →gross value. →G

baby scale
A scale with a trough-shaped →load receptor for →weighing
babies.

back-weighing
(Re)→weighing of a sample after a chemical or thermal
reaction, or physical process, that changes the mass of the
sample.

balance
→Weighing instrument, intended predominantly for medium to low capacity →weighments, with moderate to high
resolutions, mostly used indoors, often in laboratory environments and typically of OIML class ① or ②. →beam
balance, →comparator balance, →density balance, →labo-
ratory balance, →precision balance, →analytical balance

balance beam
→lever, →design and function of a mechanical balance

balance for measuring surface tension
→surface tension balance

bar code
In a bar code, numeric and alphanumeric characters are
represented as combinations of bars and spaces of different
width. The width and separation of the bars represent the
coding, which can be read and evaluated with corresponding bar code readers. Different types of code have been
developed for different applications. For →prepackages,
these are the →UPC Code (Universal Product Code, in the
USA and Canada, Fig 16a) and the →EAN code in Europe
and many countries overseas, excluding Japan, Fig. 16b).
→data matrix code

bar weight
→Weight piece in the form of a block. →OIML weight
classes. →Directive on Medium Accuracy Weights

base price
The price of a specific reference quantity of merchandise
stipulated by regulations, usually the price per kilogram.

bathroom scale
→Person scale for home use, usually located in the
bathroom.

a)

b)

Fig. 16
Bar codes
a) UPC Code;
b) EAN Code

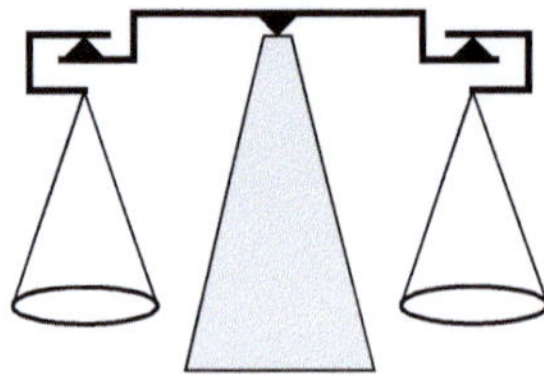

Fig. 17
Principle of the equal-arm beam balance

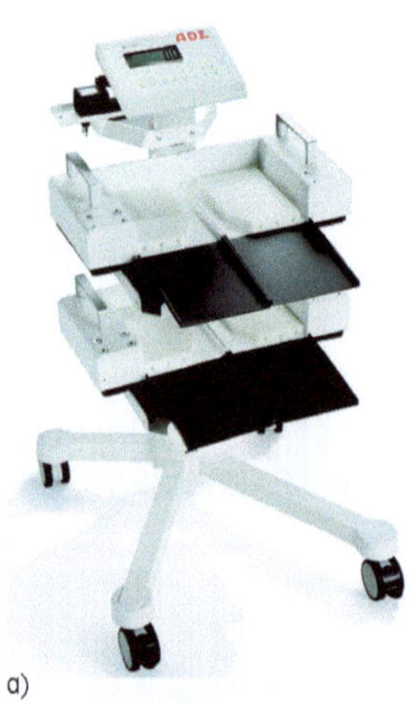

a)

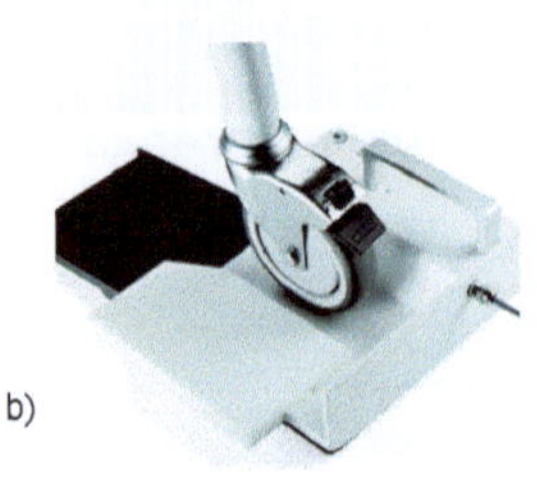

b)

Fig. 18
Bed scale
a) control unit with 4 load cells;
b) load cell under the bed leg
(Images by courtesy of Strack AG,
Schaffhausen, CH)

beam balance

A →balance in which the →load receptors are supported by →bearings or →joints at the ends of a balance beam, and where the balance beam is supported by a bearing or a joint between its ends (Fig. 17). If the beam balance has only one lever, it is referred to as a single-beam balance; if it has several levers which are connected by →links, it is called a combination beam balance. Single-beam balances are subdivided into →two-knife balances and →three-knife balances. Beam balances can be designed as equal-arm (ratio of mechanical advantage 1:1) or as unequal-arm balance (other ratios of mechanical advantage). →knife-edge bearing

beam load cell

→Spring element, Fig. 153a and b.

bearing

Depending on the type, a bearing fixes one or more degrees of freedom of translation or rotation of a mechanical component. A bearing thus transmits guiding forces and moments between the components. →knife-edge bearing, →taut band suspension

bed scale

Scale with a specially designed →load receptor capable of accommodating a hospital bed (Fig. 18). Used mainly to monitor changes in weight of a patient lying in the bed while undergoing medical treatment (for dialysis, burns, etc.). Frequently fitted with limit switches.

below-the-balance weighing

A weighment that is performed below the balance with the aid of a →hanger through the balance or a hanger around the balance as, for example, in hydrostatic weighments (→hydrostatic balance) or when weighing magnetic material. The →weighed object is placed on a →load pan below the balance. →around-balance hanger

belt loading

Loading of a conveyor belt with conveyed material; expressed as mass per unit of length (e.g. kg/m). →belt weigher

belt weigher

→conveyor belt weigher

belt-conveyor scale

→conveyor belt weigher

bench scale

A scale with a →weighing capacity of up to approximately
30 kg that is used on counters, benches, or tables, e.g.
→counter scale, →precision balance, →household scale.

Béranger scale

→Counter scale (Fig. 19a), for the principle of which Joseph
Béranger (*1802, †1870) applied for a patent [4] in 1847,
in which the two →load receptors are each supported by an
auxiliary lever in addition to the main lever (Fig. 19b and
Fig. 19c). Each →platform rests on multiple points, thereby
allowing the torque caused by →eccentric loading to be better
compensated. This makes the Béranger scale more robust
and less susceptible to oscillations than, for example, the
→Roberval scale. →Pfanzeder scale

Fig. 19a
Béranger table scale

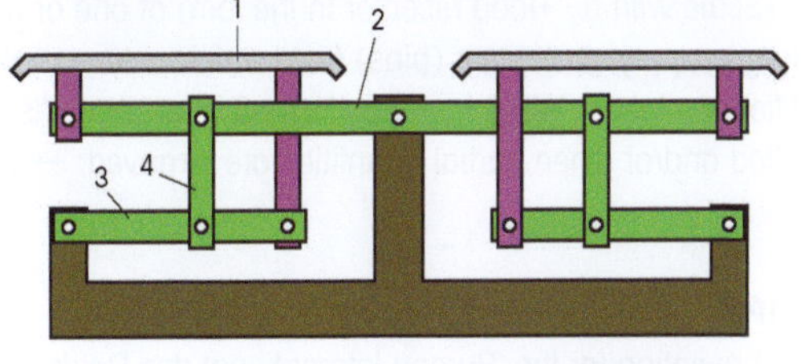

Fig. 19b
Diagram of a Béranger table scale

1: weighing pan
2: main lever
3: auxiliary lever
4: link

Fig. 19c
Lever system of a Béranger
flat-pan scale

(Images 19a and c by courtesy of
Pfunds Museum Kleinsassen/Rhön,
Hofbieber-Kleinsassen, DE)

---

[4] granted in 1849

BEV

BEV are the German initials of the Austrian Office of Metrology and Surveying, the Austrian →national metrology institute with headquarters in Vienna (www.bev.gv.at).

bias

1. The difference between the expectation [5] of the test results of an accepted reference value and that reference value.
   [ISO 5725] 3.8
2. Estimate of the systematic measurement error.
   [VIM:2008] 2.18
   →systematic error

bidirectional interface

→Interface through which an instrument acts both as data source and data receiver, in contrast to a solely input or output interface.

BIML

Abbreviation for 'Bureau International de Métrologie Légale', the International Bureau of Legal Metrology. The headquarters offices of the BIML are in Paris. Its responsibilities include the management and organization of the OIML, the preparation of →OIML recommendations and documents, and the convening of meetings, for example to approve such items (www.oiml.org/information/biml.html).

bin scale

→Scale with a →load receptor in the form of one or more large supply containers (bins) from which only partial quantities are taken. Weighing is performed when the bins are filled and/or when partial quantities are removed. →hopper scale

BIPM

Abbreviation for the 'Bureau International des Poids et Mesures', the International Bureau of Weights and Measures, which has its headquarters in Sèvres, a suburb of Paris (www.bipm.org). The task of the BIPM is to ensure worldwide uniformity of measurements and their traceability to the International System of Units (SI). The BIPM was founded for this purpose within the context of the Meter Convention, and operates under the supervision of the International Committee for Weights and Measures (→CIPM).

---

[5] what is meant here is the statistical expectation, i.e. the average of several results

Borda weighing method

→Substitution weighing named after Jean-Charles de Borda (*1733–†1799) in which the unknown mass of the →sample on the one pan is compensated on the other pan by an auxiliary tare load of approximately equal mass. The sample is then replaced with →weights of known mass (→reference mass) until the balance attains the same state (→weighing result) as with the sample. The Borda method can also be used on →single-pan balances, for example when comparing masses (→mass comparator).

Bouguer anomaly

Deviations of up to 0.01% from →standard gravity that are caused by local variations in density of the Earth's crust and mantle, named after Pierre Bouguer.

bridge

1. General name for a →load receptor with multiple supports. →weighbridge
2. Electric circuit for the measurement of electrical quantities. →measurement bridge, →Wheatstone bridge

bridge scale

General name for →scales in which the suspension (support elements) of the →load receptor is arranged underneath and hence loading is not hindered by any suspension devices installed above the load receptor. This constructional form of the load receptor is called a bridge scale, and this type of load receptor a →weighbridge. The weighing platform rests on several elements that are connected by joints (not necessarily lying in a horizontal plane) that have linked levers (Fig. 20), as for example in the →decimal balance (Fig. 38). The weighbridge can also rest directly on multiple →load cells (→road vehicle scale). Heavy-duty scales are often assembled from multi-part linked weighing platforms.

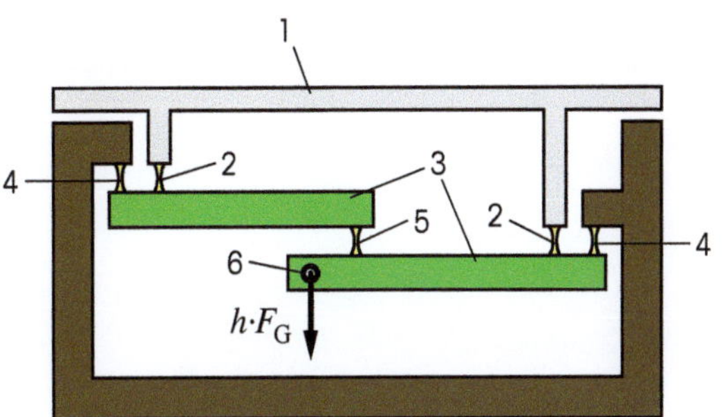

Fig. 20
Diagrammatic cross section of a bridge scale

1: load receptor (weighing platform)
2: supporting joint
3: lever
4: lever joint
5: link
6: force transfer to the load cell
F_G: weight force of load
h: leverage

bubble level
→circular level indicator

buoyancy
A force that acts on any body that is immersed in a →fluid and acts against →gravity, thereby causing an effective reduction in weight. According to Archimedes' principle, the buoyancy force F_B is equal to the →weight force of the displaced fluid

$$F_B = m_F g = (\rho_F V) g$$

where

$\quad m_F$ mass of the displaced fluid
$\quad g$ →local gravity
$\quad \rho_F$ density of the fluid
$\quad V$ volume of the body.

The residual effective weight force F_{G*} (→effective weight) of the immersed body is therefore

$$F_{G*} = F_G - F_B = mg - \left(m\frac{\rho_F}{\rho}\right)g = mg\left(1 - \frac{\rho_F}{\rho}\right)$$

where

$\quad m$ mass of the body
$\quad \rho$ density of the body
$\quad$ (other symbols as explained above).
→air buoyancy, →effective mass

buoyancy force
→buoyancy

burette
A burette is a glass tube marked with a →scale that has a tap at its lower end and is used to dispense known amounts of liquid, mainly for →titration (Fig. 21a). The volume (→volumetry) that has been dispensed can be read off the scale. There are also burettes that are integrated into a titration apparatus with supply bottle (Fig. 21b). A further version consists of a bottle-mounting instrument with a piston-cylinder system. An advantage of this version is the digitally readable volume (digital burette, Fig. 21c). →pipette

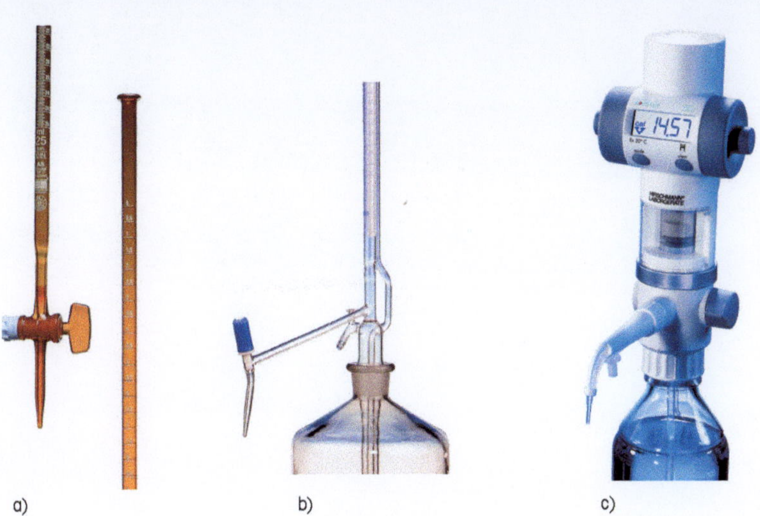

Fig. 21
a) burette;
b) Titration apparatus with supply bottle;
c) Digital burette
(Images by courtesy of Hirschmann Laboratory Instruments, Eberstadt, DE)

calibrate, to

1. Action of establishing a relation between the quantity
   values provided by measurement standards and corresponding indications ([VIM:2008] 2.39).
2. Action of determining the →deviation between the
   →measurement value and the true value of the →meas-
   urand under specified measurement conditions without
   making any changes. (compare: →adjustment)
3. Term used in non-technical language for →'to adjust',
   especially in the United States of America.

calibration

1. Result of the action of calibrating (→'to calibrate') an
   instrument.
2. Non-technical term for the test of the →sensitivity of
   a measuring instrument with the aid of a reference
   (→standard) without →sensitivity adjustment.
3. Term used in non-technical language for →'adjustment'.

calibration laboratory meeting ISO 17025

Testing or calibration laboratory accredited to ISO/IEC 17025
(→ISO 17025). The →accreditation confirms that the laboratory possesses the competence to perform tests and/or
calibrations according to the requirements of ISO/IEC 17025.

calibration service

Organization for the →accreditation and monitoring of
calibration laboratories in industrial and other institutions
(e.g. technical inspection authorities, university institutes,
national authorities) with the aim of ensuring the →trace-
ability of measuring equipment and standards to national
standards, particularly in industrial metrology. Examples are
the German Calibration Service (→DKD) or the United Kingdom Accreditation Service (UKAS).

calibration weight

→reference weight

canister load cell

→pin load cell

carat scale

A →balance that indicates the →weighing value in →metric
carats (ct) and is particularly suitable for weighing precious
stones (Fig. 22).

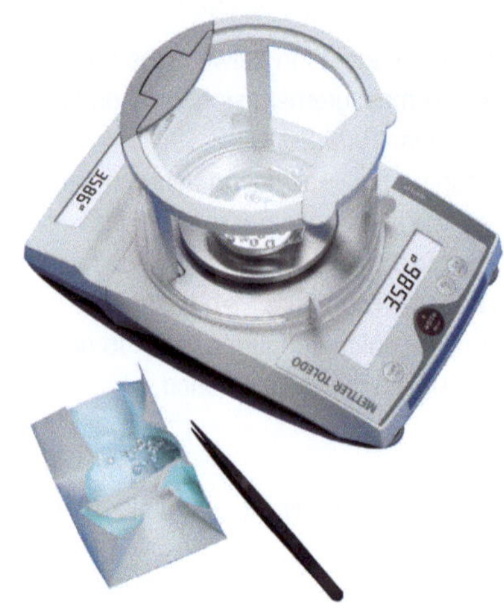

Fig. 22
Carat scale

carat, metric
→metric carat

cash register systems
Electronically programmable cash registers connected by non-interacting data interfaces to →non-automatic weighing instruments in public points of sale.

catch weigher
Automatic instrument for single weighments (→automatic weighing instrument).

CE mark
By affixing the CE mark (Fig. 23), the manufacturer declares that the marked instrument conforms to all applicable European Directives. For weighing instruments, these are the →Low Voltage Directive 2006/95/EC, →EMC Directive 2004/108/EC, →Directive on Non-Automatic Weighing Instruments 2009/23/EC (where applicable), →ATEX 95 Directive 94/9/EC (where applicable), and Measuring Instruments Directive 2004/22/EC (where applicable). The additional affixation of the so-called →Green M indicates specific conformity of a weighing instrument with the →Directive on Non-Automatic Weighing Instruments 2009/23/EC or the →Measuring Instruments Directive 2004/22/EC. In addition to the CE mark on the instrument, the manufacturer issues an →EC Declaration of Conformity, in which conformity with all applicable European Directives is explicitly confirmed.

CE marking for EC verification
CE marking for →EC verification comprises the →CE mark, the last two digits of the year number in which the verifica-

Fig. 23
CE mark

tion was performed (→initial verification), and the identification number of the →Notified Body that performed the →EC verification, or under whose supervision the EC verification was performed by the manufacturer (Fig. 24). The CE marking is applied by the manufacturer.

CE year notation

The CE year notation consists of the last two digits of the year in which the →initial verification took place. →CE marking for EC verification, →EC verification marking, →stamping mark

center of gravity

The point of a rigid body around which no torque occurs in a homogeneous force field (e.g. gravitational field at the surface of the Earth). It is therefore also the point at which the mass of a body or system can be imagined as being concentrated. →equilibrium position

certificate of conformity

The conformance of verified weighing instruments to the →EC type approval is documented by the verification authority (→Notified Body) by issuing a certificate of conformity at the time of verification.

certified computer

A computer as a component or →auxiliary device of a verifiable measuring instrument in which important certified functions are implemented in certified programs. →Hardware and →software protective measures guarantee the security of the certified programs and allow the separation of non-certified program parts. The goal of the protective measures is to achieve security of the certified programs at the same time as free programmability of the user programs. →legally relevant software

certified PC

→certified computer

characteristic curve

The relationship between the input variable and output variable of a measuring instrument (Fig. 25). The characteristic curve is obtained by recording, and usually displaying graphically, the output values for all possible values of the input range.

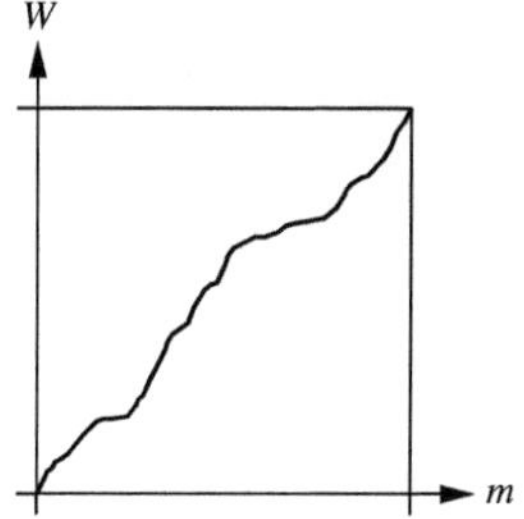

Fig. 25
Characteristic curve of load m
and weighing value W
of a weighing instrument
(deviations shown enlarged)

$\mathsf{C}\,\mathsf{E}\,\mathbf{99}$

0104

Fig. 24
CE marking for EC verification

characteristic curve of a load cell
→Characteristic curve relating the →load (input variable)
and the output variable (e.g. voltage, pressure of a liquid) of
a →load cell.

characteristic curve of a weighing instrument
→Characteristic curve relating the →load (input variable)
to the →indication (output variable) of a →weighing instrument.

checkout scale
→Scale for →public point of sale (checkout) installed at
the exit of a supermarket, usually with price calculation and
connection to the cash register. →counter scale

checkweigher
A balance used to check
1. quantities that are separated by mass or volume, or
2. automatically functioning balances or filling machines for
   the reweighing of prepackages.
→automatic checkweigher

CIPM
Abbreviation for 'Comité International des poids et mesures',
the International Committee for Weights and Measures. The
CIPM is made up of eighteen individuals, each from a different Member State under the Metre Convention. Its principal
task is to promote world-wide uniformity in units of measurement (→International System of Units) by direct action
or by submitting draft resolutions to the General Conference
(CGPM, General Conference on Weights and Measures).
(www.bipm.org/en/committees/cipm) →BIPM

circular level indicator
→Level indicator, the liquid of which is enclosed in a circular container (Fig. 26), a.k.a. spirit level or bubble level.

classify according to mass, to
Action of determining the affiliation of similar objects to
specified classes according to their mass without separating
them from each other (as opposed to →to sort).

coarse dispensing
→fine dispensing

coarse display
→available capacity indicator

Fig. 26
Circular level indicator

coarse feed

In →automatic weighing instruments, material flowing into the →load receptor for an approximate apportionment (→to apportion) of the fill quantity. The coarse feed is followed by the →fine feed. →fine dispensing

coarse range

→normal range, →fine range

coarse weighing

Approximate apportioning (→to apportion) of a sample as a first step of a →weighed-in quantity.

coefficient of variation

→variation coefficient

combination scale

A combination →scale is used to fill packages optimally to a desired target weight. It comprises a number of individual weighing stations (weigh hoppers), each holding just a fraction of the package weight, arranged around a central supply of the product being weighed (Fig. 27a). Each weigh hopper is filled and weighed. From the fill quantities, which in terms of weight are randomly distributed, a computer determines the combination that comes closest to the target weight. These weigh hoppers are then selected and their content is filled together into the next package (Fig. 27b).

combined error

A →measurement deviation that is composed of several random or systematic components. Known systematic measurement deviations (→systematic error) must be taken into account (correction). All other components (→random errors and →measurement uncertainty of the systematic errors) are considered to be random quantities when the measurement uncertainty is determined. →uncertainty

combined rail car and road vehicle scale

→Scale that can be used to weigh both rail cars and road vehicles. →automatic rail scale, →road vehicle scale

commercial scale

→weighing instrument of medium accuracy

commercial weight

1. Non-technical term for weights that are traditionally used in handling merchandise. Commercial weights exist with nominations of 125 g, 250 g, 500 g, 1 kg, 2 kg, 10 kg,

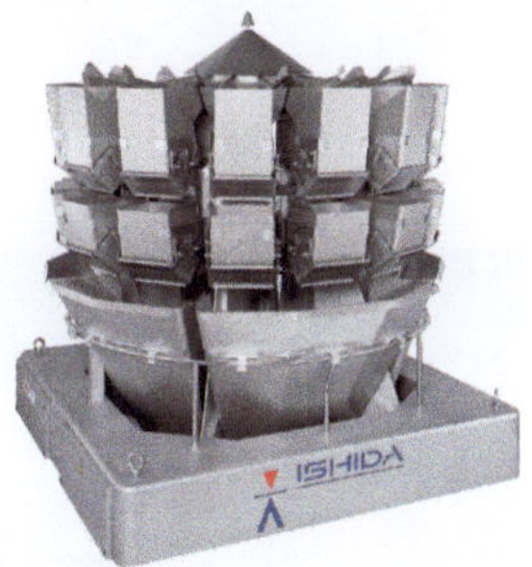

Fig. 27a
Combination scale

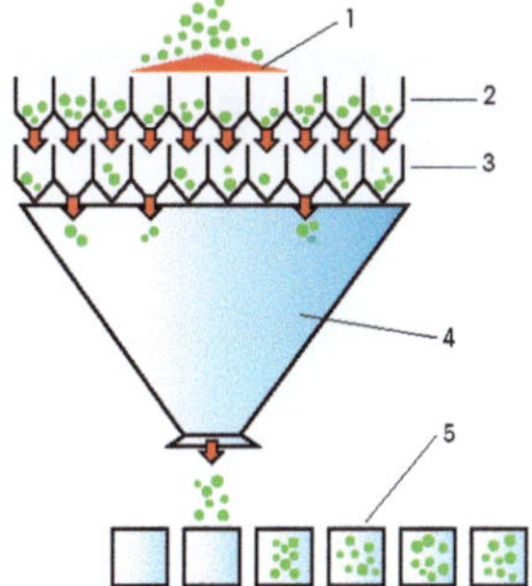

Fig. 27b
Operating principle
of a combination scale

1: dispersion feeder
2: pool hoppers
3: weigh hoppers
4: discharge chute
5: packages

(Images by courtesy of Ishida Europe, Birmingham, UK)

25 kg, 50 kg, etc., which with the introduction of the →OIML weight classes were replaced by the 1-2-5 pieces.
2. Term defined in the German →Verification Ordinance for →weight pieces of class OIML M3 (→OIML weight classes) ([VO] Appendix 8, Section 2, 2.1).

comparator balance
→mass comparator

compensation coil
The moving coil in the permanent magnet system of an →electrodynamic converter as used for →electromagnetic force compensation (Fig. 28).

compensation current
Current that flows through the →compensation coil of an →electrodynamic converter to produce the compensation force.

a)

b)

Fig. 28
Compensation coil
a) winding on the coil carrier;
b) completely installed coil

compensation principle
→force compensation

compression column load cell
→pin load cell

compression weighing cell
→Spring element, Fig. 153c, →pin load cell.

compulsory verification
According to the →Weights and Measures Act, this obligation exists for certain instruments (including →weighing instruments and →weight pieces) if they are used in commercial and official activities, in the field of public health, or in the preparation and testing of pharmaceutical products.

computer, certified
→certified computer

confidence interval
→coverage interval

confidence level
Probability that the expected value of a →measurand lies within the (usually symmetrical) →coverage interval $\pm U$ of the →measurement value. For a given coverage interval, the confidence level depends on the probability distribution of the measurement value.

configuration

→method

connecting hanger

A movable connecting link between the suspended
→load receptor or weight receiver and the corresponding
→hanger.

connecting lever

In composite weighing instruments, →levers that are connected between the load lever and the weighing lever.

constructional requirements

Legal regulations, guidelines, or standards that apply to
→weighing instruments. For verified weighing instruments
these may include suitability, safe operation, display of the
weighing value, etc.

control chart

In the control chart, values that were obtained from repeated
sample checks of a process are entered, usually graphically.
The control chart is used to monitor the process. Should a
sample value attain the →warning limit or →control limit,
the process must be corrected if necessary.

control limit

1. →Tolerance of a process relative to its target value.
   Violation of the tolerance is an infringement of the quality
   requirements, and therefore requires a correction of the
   process. →warning limit
2. A term used in →prepackage process control to designate upper and lower weight limits that a package must
   not violate. →minus deviation

control unit

→Electronic device in a weighing instrument that performs
the analog-digital conversion (→analog-digital converter) of
the →output signal of the →load cell, as well as the further
processing of these data including display of the →weighing
result in →units of mass. ([OIML R 76-1] 2.2.2)

conventional mass

→Weighing value that is obtained when a weighment is
performed under the conditions specified in OIML D 28, i.e.
in air of density $\rho_0 = 1.2$ kg/m^3, at a temperature of 20 °C,
and using a reference standard of density $\rho_c = 8000$ kg/m^3
([OIML D 28], [DIN 1305] 4). This weighing value is

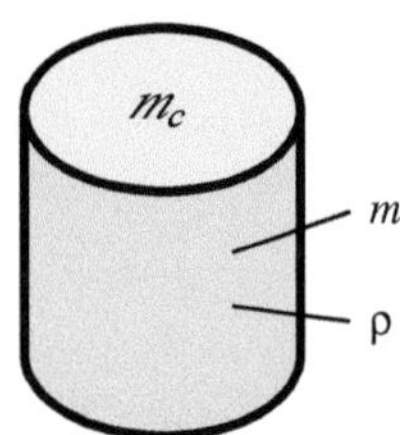

Fig. 29
Weight piece with indication of its
conventional mass

m: mass
ρ: density
m_c: conventional mass
(mass value assigned according to
convention OIML D 28)

assigned to a body (in particular to a →weight piece) of
mass m with density ρ as conventional mass m_c (Fig. 29)

$$m_c = \frac{1 - \dfrac{\rho_0}{\rho}}{1 - \dfrac{\rho_0}{\rho_c}}\, m = \frac{1 - \dfrac{\rho_0}{\rho}}{0.99985}\, m\,.$$

The rationale for the convention is that when the mass unit
is passed on in the form of weights that are usually made
of stainless steel with a density of around 8000 kg/m^3, the
→correction for air buoyancy can be largely ignored (except
for the highest →accuracy classes).
The conventional mass of a body depends only on its mass
and density, since the other quantities (density of air and
density of the reference) are defined by the convention. The
mass m of a body can thus be derived from the conventional
mass m_c at any time according to the equation

$$m = \frac{1 - \dfrac{\rho_0}{\rho_c}}{1 - \dfrac{\rho_0}{\rho}}\, m_c = \frac{0.99985}{1 - \dfrac{\rho_0}{\rho}}\, m_c\,.$$

A weighing instrument whose →sensitivity was adjusted
with a →reference weight of density 8000 kg/m^3 shows the
→conventional mass of the weighed object provided that
the weighing is performed at the conventional air density of
1.2 kg/m^3 and a temperature of 20°C.

conventional scale interval
→Actual scale interval expressed in units of mass and fixed
by directives that is used to allocate weighing instruments
not equipped with indicating devices to their appropriate
weight class.

conventional value
→conventional mass, [OIML D 28]

conversion factor
→electrodynamic converter, →unit conversion factor

converter
Device that converts quantities of one medium into quantities of another medium, the converter creating a relationship
between the quantities (usually proportionality). Media can
be, for example, physical or numerical quantities. Examples
of converters are the →electrodynamic converter and the
→analog-digital converter.

conveyor belt weigher
A →scale that determines the weight of unpacked, loose,

continuously transported material by weighing the →belt loading and measuring the belt speed while the material is being transported (integrated scale) (Fig. 30). Conveyor belt weighers can also be designed as adding scales that weigh the material in sections.

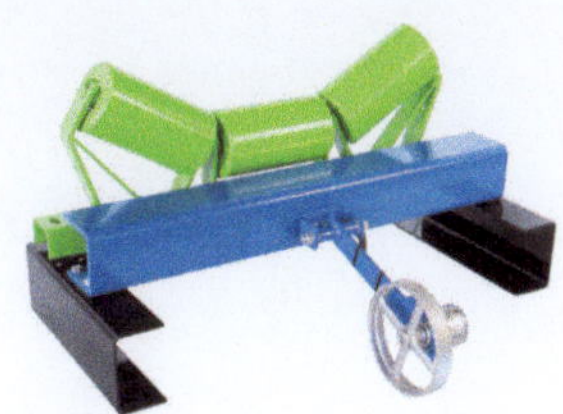

Fig. 30
Conveyor belt weigher
(Image by courtesy of Pfreunt GmbH,
Südlohn, DE)

Coriolis mass counter

In the Coriolis →mass counter, the liquid being measured flows through the instrument either through two parallel measuring tubes of the same construction or through a double expansion loop (Fig. 31). The measuring tubes are made to vibrate by, for example, field coils. When a liquid flows through the vibrating tube, Coriolis forces arise that influence the normal vibration. These forces cause the measuring tube to experience an angular deflection, from which the mass flow dm/dt can be measured.

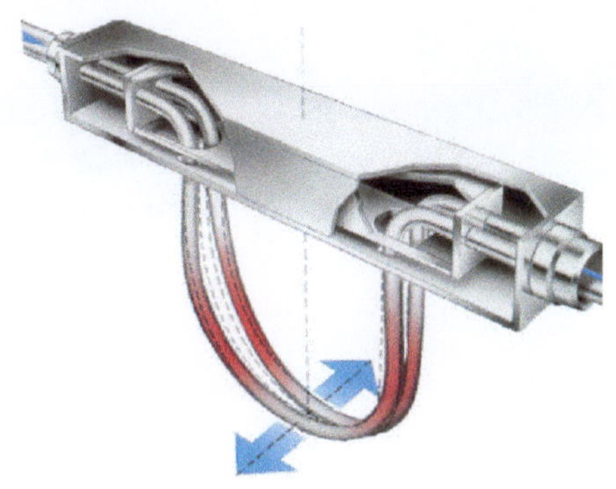

Fig. 31
Coriolis mass counter
(Image by courtesy of Helios &
Zaschel GmbH, Mühltal, DE)

correction for air buoyancy
→air buoyancy correction

counter
A device used to determine increasing piece numbers, parts of units, e.g. length, impulses, etc.

counter scale
A verifiable →scale at a public point of sale that indicates the weight, base price, and purchase price, usually in the form of a →deflection balance or →electromechanical weighing instrument (Fig. 32). The display must always be visible to both buyer and seller. If this is not possible with a single display, a double-sided display must be used ([OIML R 76-1] 4.13.6).

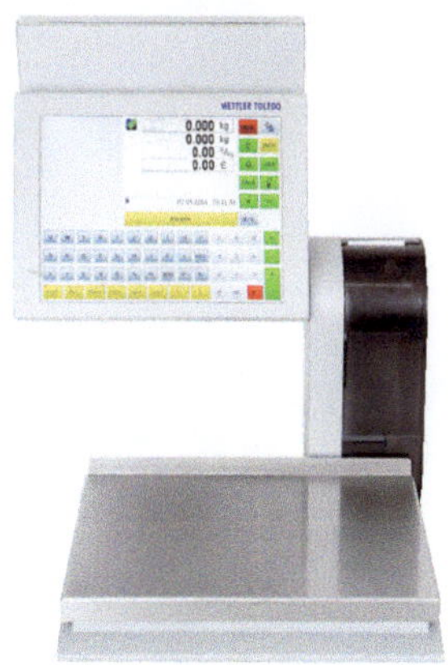

Fig. 32
Counter scale

counterpoise weight
Common term in the United States of America for →rider.

counting device

A device for totaling pieces, fill quantities, etc.

counting scale

A →scale with special equipment to count numbers of pieces of identical weight (Fig. 33). Electronic counting scales determine the mean individual weight and total weight of the pieces to be counted in separate operations and calculate the number of pieces by division. Mechanical counting scales are modified →decimal balances, centesimal, or →sliding weight balances with fixed or variable →ratios of mechanical advantage.

Fig. 33
Counting scale

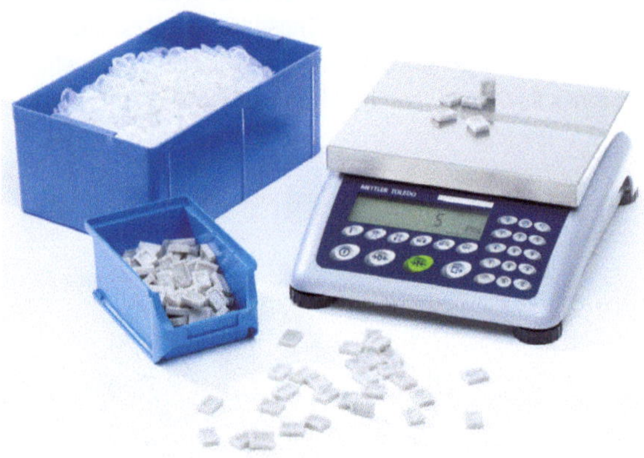

coverage factor (k)

→expansion factor

coverage interval

Range U (→uncertainty interval) on both sides of a →measurement value (corrected for →systematic error) that contains the expected value of the →measurand with a specified →confidence level. Assuming that the underlying quantity is normally distributed (→normal distribution), the confidence level for $k = 1$ (→expansion factor) is approximately 68%, for $k = 2$ approximately 95%, and for $k = 3$ approximately 99.7%.

crane scale

A →scale used to weigh the load suspended from a crane. The scale can either be suspended from the crane hook (Fig. 34) or integrated in the structure of the crane. In the latter case the cable drum and drive, as well as all cable guide parts, are part of the scale preload. (compare: →rope-tension scale)

Fig. 34
Crane scale
(Image by courtesy of Dini Argeo,
Spezzano di Fiorano, Modena, IT)

creep error
The deviation that arises when a drifting (→drift) →measurement value is read, printed, or processed before the gradual approach to the stable state (e.g. of the measurement value) has been completed. →settling

cross-flexed bearing
→cross-flexed spring joint

cross-flexed spring joint
Two metal bands that are positioned adjacent to each other with an imaginary common fulcrum (pivot) and whose planes are at right angles to each other (Fig. 35). →joint, →flexible joint

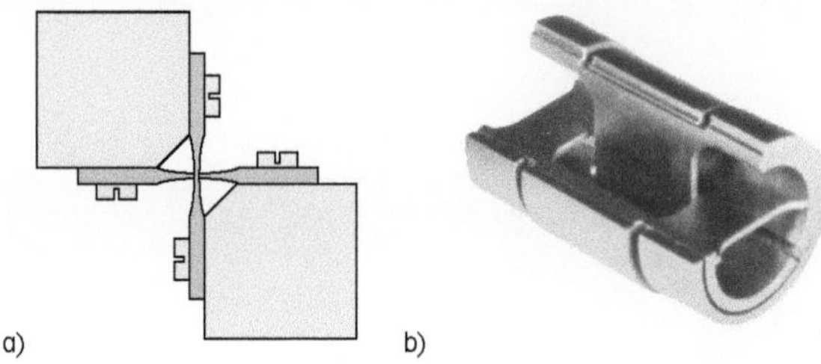

Fig. 35
Cross-flexed spring joint
a) operating principle;
b) exemplary embodiment

(Image 35b by courtesy of GAD Elektronik-Komponenten Vertriebs GmbH, Nussloch, DE)

ct
Unit symbol for the →metric carat.

current balance
Measuring device, which is no longer in use, with a high-resolution weighing instrument for displaying the SI base unit ampere (→International System of Units).

customer keys
Input keypad of a →counter scale that is located on the customer side of the scale. Input keypads are only permitted on →self-service weighing instruments. (compare: →salesperson keys)

cylindrical weight
→Weight piece in the form of a cylinder. →OIML weight classes, →Directive on Medium Accuracy Weights

d
→scale interval, →digit

D/A converter
→digital-analog converter

damping
Reduction of the amplitude of a periodically varying quantity. In the case of a weighing instrument, the reduction in amplitude of oscillations until a stable equilibrium is reached (→settling). The oscillation energy is directly dissipated, e.g. as frictional heat or electrical heat, or transformed into a different form (e.g. electrical energy in the case of electrodynamically compensating weighing instruments). →design and function of an electrodynamic balance, →damping systems

damping device
Device for →damping the oscillating moving parts of a weighing instrument by means of suitable →damping systems. The device causes the indicating component to attain its →equilibrium position faster.

damping systems
To avoid affecting the →repeatability of a →weighing instrument (or →load cell), →damping devices are used in which the →damping is proportional to the speed of movement (so-called viscous friction).
1. Air damping
   A damping plate connected to the moving part that alternately compresses the air in the upper and lower chamber of the damping cylinder. The kinetic energy is thereby converted into heat of compression and friction of the air (Fig. 36a).
2. Liquid damping
   The kinetic energy is converted into frictional heat through the friction of the liquid (usually oil) (Fig. 36b).
3. Eddy current damping
   A non-magnetic electrically conductive (copper, aluminum) damping vane is rigidly attached to the oscillating part and moves between the poles of a magnet. The resulting eddy currents convert the kinetic energy into heat (Fig. 36c).

In load cells that are equipped with a controller and a measurement →converter that exchanges power (e.g. electrodynamic compensation), damping of the complete system can be achieved by the selection of suitable control parameters.

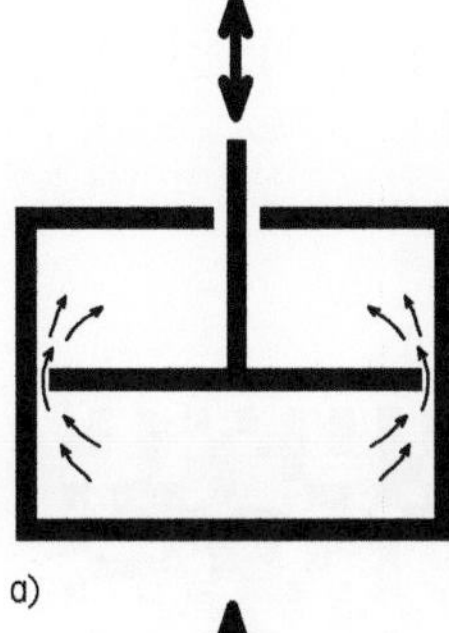

a)

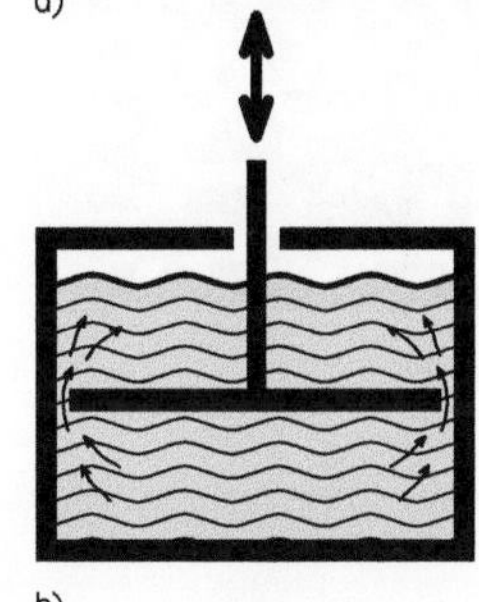

b)

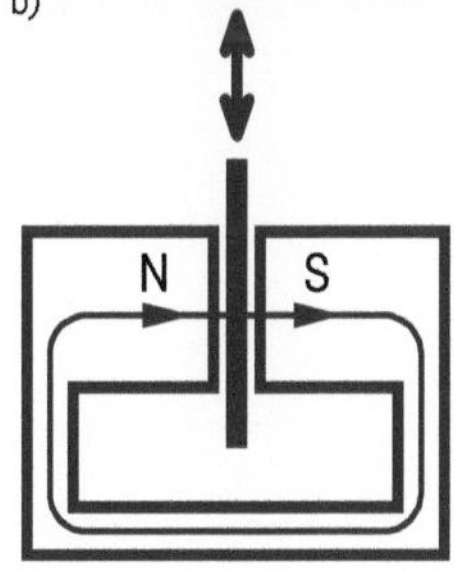

c)

Fig. 36
Damping systems
a) air damping;
b) fluid damping;
c) eddy current damping

N: magnetic north pole
S: magn. south pole

data bus

Multipoint electric connection (→interface) between multiple participants with a common medium (e.g. electromagnetic via cables, wireless, or optical). A wired data bus has a long main cable to which the participants are connected via short spur lines. The bus structure is thus a different form of multipoint connection than a star or ring structure. A databus can be used to connect a PC with peripherals and measuring instruments, or to connect modules inside the computer, e.g. an address and data bus to connect the processor memory to peripheral components.

data concentrator

Central control and storage unit into which measured values and data from several →weighing instruments or instruments flow, are temporarily stored, and when required passed on to an output device, e.g. a printing unit.

Fig. 37
Matrix code

data matrix code

Representation of numeric and alphanumeric characters by means of a pattern of square dots in a square surface (Fig. 37). A square with a relatively small number of subdivisions can already encode an enormous amount of data. For example, a 10×10 matrix code can encode $2^{100} \approx 10^{30}$ bits of data. This allows sufficient redundancy to correct any errors that may occur. →bar code

data memory

Memory built into a weighing instrument and executed as a pure software solution in an instrument, or as an external auxiliary device for long-term storage of verified weighing results with all data that are associated with a weighing operation and important for storage.
([OIML R 76-1] 2.8.1)

data plate

Plate on which the designations and →inscriptions stipulated in the verification, safety, and other regulations are all displayed.
1. Plate containing information for a more detailed identification of the respective product, for example, name of the manufacturer, model, serial number, maximum capacity, operating voltage, power supply frequency, type data, approval data, prohibition of use at public points of sale, instructions regarding intended use, safety instructions, etc. The required →inscriptions are defined more specifically in the applicable legal regulations for the respective

instrument (e.g. →Directive on Non-Automatic Weighing
Instruments, →ATEX 95 Directive).
2. More specifically, the term "marking" is used for the →CE
   mark (CE marking), while the remaining items of information on the plate are referred to as →'inscriptions'.
   In non-technical language, the terms 'name plate' and
   'type plate' are also used synonymously for 'data plate'.

data storage device

Non-technical expression for a →verifiable memory device
in a certified →weighing system to which non-certified auxiliary devices or data processing systems with printer are
connected.

data transmission

Transformation of information by means of electrical signals
over electric conductors or wirelessly between two instruments, e.g. a balance and a printer or PC.

dead load

Sum of the mass of all elements connected to the movable
part of the mechanical system of a weighing instrument. If
the instrument contains →levers or →parallel guides, the
masses of these elements have to be considered according
to their effect.

decimal balance

Unequal-arm balance in which an arrangement of levers
reduces the effect of the load by a factor of ten, so that
→weight pieces with a mass of one tenth of the mass of the
load are sufficient to compensate the load (Fig. 38). The
decimal balance is usually executed as a →bridge scale.

Fig. 38
Decimal balance
(Image by courtesy of Pfunds
Museum Kleinsassen/Rhön,
Hofbieber-Kleinsassen, DE)

Fig. 39
Deflection balance with parallel guide

declaration of compatibility
Evidence verified and declared by the manufacturer confirming the compatibility of the →modules listed in a →type approval.

declaration of conformity
→EC Declaration of Conformity

deflection balance
A →balance in which →load compensation is effected by a →deflection weighing device (Fig. 39). →physical weighing principle

deflection weighing device
→Weighing-out device of a lever balance comprising the deflection lever and the associated indicating device.
→Load compensation is effected by deflection of the lever.
→physical weighing principle

degree of protection (IP)
→degrees of protection provided by enclosures

degrees of protection provided by enclosures
Characteristics defined in standards relating to equipment safety, a.k.a. "ingress protection". IEC 60529 is concerned with the protection of electrical equipment by housings, covers, and the like. The standard concerns the protection of persons against contact with electrically live or moving parts, the protection of equipment against the ingress of solid objects and water, and defines the codes for internationally agreed types and levels of protection (IP Code). The first digit of the IP Code defines the level of protection against contact and the ingress of solid objects (Tab. 2a), the second digit defines the level of protection against ingress by water (Tab. 2b).

Tab. 2a
Level of protection: 1. digit

First digit	Protection against contact and ingress of solid objects
0	non-protected
1	≥ 50 mm diameter
2	≥ 12.5 mm diameter
3	≥ 2.5 mm diameter
4	≥ 1.0 mm diameter
5	dust-protected
6	dust-tight

Second digit	Protection against water
0	non-protected
1	vertically dripping
2	dripping (15° tilted)
3	spraying
4	splashing
5	jetting
6	powerful jetting
7	temporary immersion
8	continuous immersion

Tab. 2b
Level of protection: 2. digit

Delta Range balance

Weighing instrument with a movable →fine range. →multi-interval instrument

DeltaRange® (DR)

Second range of a weighing instrument that usually has a ten times smaller →readability (→fine range) than the →normal range. The fine range can be called up at a keystroke over the entire weighing range although it covers only part of the normal range. →multi-interval instrument. (compare: →Dual Range)

DeltaTrac®

Quasi-analog display consisting of segments arranged in a circle that provides a visual complement to a digital display. Acting as a gross indicator, for example, the number of filled segments provides information as to how much of the weighing range has already been taken up by the load (Fig. 40a). When used as a →weighing-in aid (target weight) or for →differential weighing, one segment functions as coarse display, and a second as fine display, which on reaching the target weight comes to rest between two tolerance marks (Fig. 40b).

denier

Unit of measure for the fineness of yarn (→yarn count): 1 den = 1 g/9 km.

denier balance

Former term for a precision balance to determine yarn fineness (→yarn count). →yarn balance

densitometer

A measuring instrument to determine the →density of fluids (Fig. 41). A U-shaped hollow glass tube is caused to

---

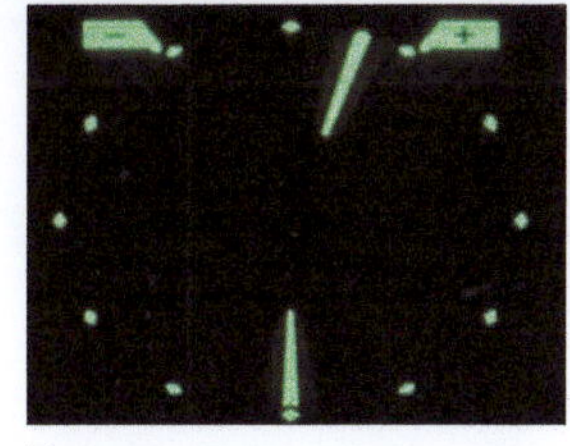

Fig. 40
Semi-analog graphic display
(DeltaTrac)
a) as graphic display;
b) as weighing-in aid

vibrate. When the glass tube is filled with a sample, the frequency of vibration decreases. The density of the sample can be determined from the frequency of vibration. →density determination 2.4

Fig. 41
Densitometer with vibrating U-tube

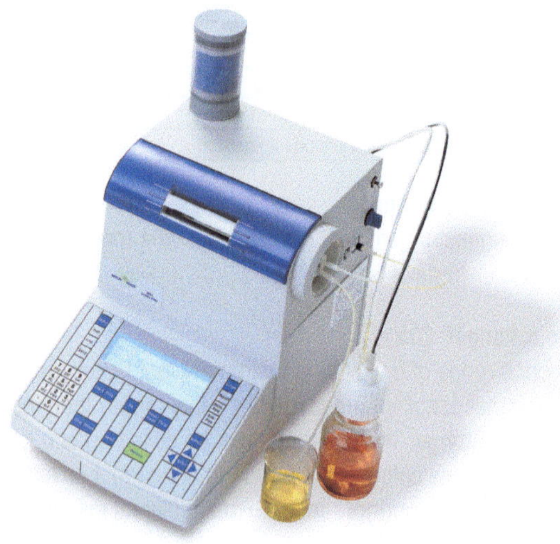

density

The density ρ of a body is the ratio of its →mass m to its →volume V

$$\rho = \frac{m}{V}$$

density balance

→hydrostatic balance

density determination

The →density of a substance can be measured directly (→densitometer) or by means of a mass determination and a →volume determination. While the mass can be determined by weighing, the volume can be determined in a number of different ways:

1. Density determination of solid bodies
   A comprehensive discussion of the density determination of solid bodies is contained in [OIML R 111-1], B.7 Density.
1.1 Bodies with geometrically simple forms can be measured and the volume calculated from the resulting dimensions (stereometry). ([OIML R 111-1] B.7.8 Test method E)
1.2 The body is submersed in a calibrated →measuring container that is filled with liquid; the volume can be read off directly on the container as the increase in the volume of liquid.

1.3 Instead of measuring the volume of liquid directly, it
can also be determined by weighing ($\rightarrow$pycnometer).
The body is first weighed (W_B). The pycnometer is
then filled with a liquid of known density ρ_L and also
weighed(W_P). The body is now placed in the pycnometer (as a result of which liquid is displaced) and
weighed again (W_{PB}). The volume of the body (V_B) is
given by the relationship

$$V_B = \frac{1}{\rho_L}(W_P + W_B - W_{PB})$$

and its density

$$\rho_B = \rho_L \frac{W_B}{W_P + W_B - W_{PB}}$$

The pycnometric method is more accurate than direct
volume measurement and is recommended for routine
density determinations of medium accuracy. (OIML
R 111-1, B.7.7 Test Method D)

1.4 According to Archimedes' Principle, the volume of a
body can be determined by successive weighments
in two mediums of different density, e.g. air and water
($\rightarrow$hydrostatic balance) (Fig. 42). The hydrostatic
method is more exact than the method of density determination described above. The volume results from the
formula

$$V_B = \frac{1}{\rho_L}(W_a - W_L)$$

and the density from

$$\rho_B = \rho_L \frac{W_a}{W_a - W_L}$$

where
V_B the volume of the sample that is to be determined
W_a the weighing value of the sample in air
W_L the weighing value of the sample in the liquid
     (e.g. water)
ρ_L the density of the liquid (e.g. water, $\rightarrow$water
     density)
Note: If the body has a low density and/or the density
must be determined more accurately, the influence of
the density of the air during weighing must be taken
into account. The following formulas then apply:
For the volume

$$V_B = \frac{W_a - W_L}{\rho_L - \rho_a}\left(1 - \frac{\rho_a}{\rho_c}\right)$$

ρ_c = 8000 kg/m^3 conventional density for the reference
     normal
ρ_a density of air ($\rightarrow$air density)

and for the density of the body:

$$\rho_B = \frac{\rho_L W_a - \rho_a W_L}{W_a - W_L}$$

([OIML R 111-1] B.70.5 Test method B)

1.5 Volume comparison

The body whose density is to be determined is compared with a body whose density is nominally identical and known (reference body). For this purpose, both bodies are compared in a →volume comparator in air as well as in a liquid (hydrostatic comparison). From the results of both comparisons, the density of the test body can be calculated. (OIML R 111-1, B.7.4 Test method A)

2.   Density determination of liquids

2.1 Volume determination by filling the liquid into a calibrated container (→volumetric flask, →measurement cylinder, →pycnometer). Mass determination by weighing before and after filling.

2.2 Determination of the density with the aid of a →hydrometer.

2.3 According to the buoyancy method on a →hydrostatic balance: A →displacement body of known volume is weighed in air and in the liquid that is being investigated (cf. 1.3) (Fig. 42). The density is given by the formula

$$\rho_L = \frac{1}{V_B} \left(W_a - W_L \right)$$

ρ_L the density of the liquid that is to be determined

W_a the weighing value of the displacement body in air

W_L the weighing value of the displacement body in the liquid (e.g. water)

V_B the volume of the displacement body.

Note: If the body has a low density and/or the density must be determined more accurately, the density of the air during the weighment in air must be taken into account. The following formula must then be used:

$$\rho_L = \frac{W_a - W_L}{V_B} \left(1 - \frac{\rho_a}{\rho_c} \right) + \rho_a$$

ρ_a density of air (→air density)

2.4 Determination of the density with a →densitometer according to the principle of the vibrating U tube.

3.   Density determination of gases

3.1 As in 2.1.

3.2 Determination of the density with two →displacement bodies of different density. For this purpose, a mass comparison is performed in the gas with two bodies whose masses and volumes are known (m_1, m_2, and V_1, V_2, respectively). From the difference between the weighing values

$$\Delta W = W_2 - W_1$$

the density of the gas can be determined

$$\rho_G = \frac{m_2 - m_1 - \Delta W}{V_2 - V_1 - \dfrac{\Delta W}{\rho_c}}$$

ρ_G the density of the gas that is to be determined
ρ_c = 8000 kg/m^3 conventional density for the reference
    normal

density determination set
The utensils required for →density determination on a balance such as containers, wire basket, →sinker, and →liquid thermometer (Fig. 42).

density of air
→air density

density of water
→water density

descriptive markings
→inscriptions

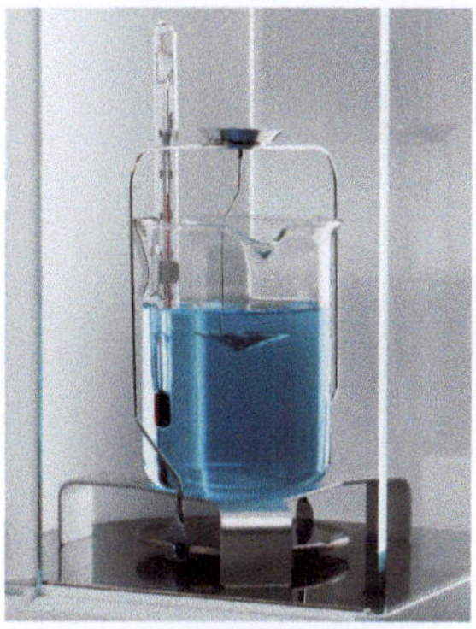

Fig. 42
Set for the determination of density

design and function of a mechanical balance
A mechanical →balance usually comprises a balance beam, one end of which is linked to a →load receptor, and the other end of which is linked to a further carrier for →weight pieces or a fixed counterweight. The lever and the load receptors are supported by →pivot joints, for example by →knife-edge bearings. The weight force of the sample to be weighed is compensated by mass comparison (→three-knife balance). The →inclination of the lever can be read off a →display device that has a →pointer and →scale, or from a →projected scale. The →measurement value is composed of the sum of the necessary counterweights or substitution weights and the readout.
There are essentially two types of mechanical balances (see also →physical weighing principle):
1. Balances with a two-arm lever and two weighing pans (e.g. →equal-arm beam balance, →sliding weight balance or →deflection balance); the →weight force of the load is compensated by weight forces of loose or built-in weight pieces at the opposite end.

2. Weighing instruments with a single-arm lever and one weighing pan (e.g. →substitution balance); the →weight force of the load is substituted by →weight pieces that are built-in at the load end (Fig. 43).

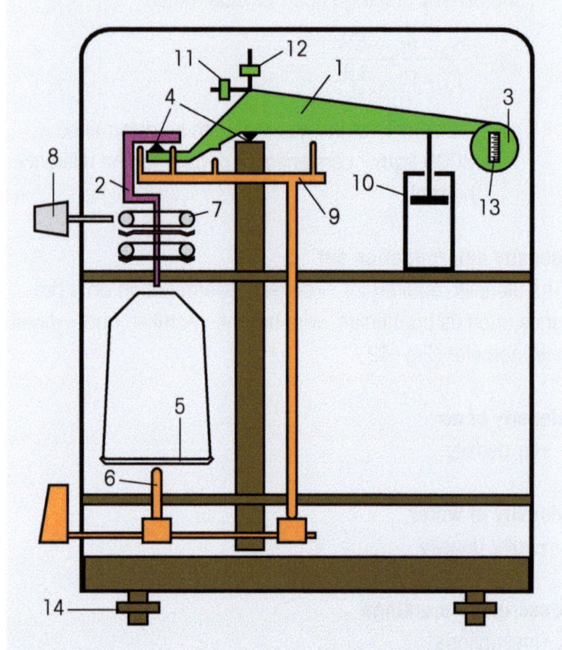

Fig. 43
Diagrammatic cross section of substitution balance as an example of a mechanical balance

1: balance beam
2: hanger
3: counterweight
4: knife-edge bearing
5: weighing pan
6: pan brake
7: dial weight
8: weight dial
9: locking device
10: air damper
11: zero-point adjuster
12: sensitivity adjuster
13: graduated plate
14: leveling screw

design and function of an electrodynamic balance

The design and function will be described by reference to the example of a →top-loading precision balance according to the principle of electromagnetic force compensation (→electrodynamic converter, →EMFC weighing instrument) (Fig. 44):

The weight force of the sample to be weighed on the weighing pan (1) passes to the hanger (2)[6]. The hanger is constrained by →guides (3) that are connected via →flexible joints (4) (→parallel guide). The link (5), which also takes the form of a flexible element, transfers the weight force to the load arm of the lever (6) that is supported at the fulcrum by flexible joints (7). The other end of the lever (force arm) holds the →compensation coil (8) of the electrodynamic converter which is located in the magnetic flux (10) of the magnet system that is created by a permanent magnet (9).

---

[6] The hanger has taken its name from the →substitution balance, although with the electrodynamic balance no weights are hung.

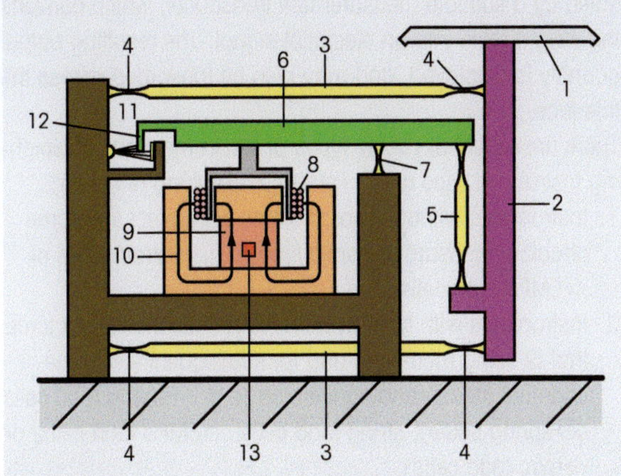

Fig. 44
Diagrammatic cross section of an
electrodynamically compensating
balance as an example of a precision
balance

 1: weighing pan
 2: hanger
 3: guide
 4: flexible joints
 5: coupling
 6: lever
 7: lever bearing
 8: compensation coil
 9: permanent magnet
10: magnetic flux
11: optical position sensor
12: position vane
13: temperature sensor

If a sample is placed on the weighing pan, the lever tilts
due to the →weight force acting upon it. The displacement
of the →position vane (12) that is registered by an electro-
optical →position sensor (11) is passed on to an electronic
controller that increases the →compensation current flowing
through the compensation coil until the lever has returned
to its original equilibrium position. Since the weight force
is proportional to the compensation force (→mechanical
advantage of the lever) and this in turn is proportional to the
compensation current (electrodynamic converter), the compensation current is also proportional to the weight force and
therefore to the load on the balance.

The →analog-digital converter converts the (analog)
compensation current into a digital quantity with the result
that the signal converter provides a digital equivalent of
the →measurand. Since the magnetic flux depends on the
temperature, the latter is measured by a temperature sensor (13) whose signal is also made available to the signal
processor (Fig. 45) which compensates for any drift of the
measurement signal caused by temperature fluctuations.
The measurement signal is linearized, translated into a
mass unit, and finally indicated, or transmitted over an
→interface.

Fig. 45
Temperature sensor
The temperature sensor is located in
an aluminum capsule in the center of
the cutaway permanent magnet.

design and function of an electromechanical weighing instrument

Electromechanical weighing instruments (often also referred
to as electronic weighing instruments) generally consist
of a →load receptor, an electromechanical →converter as
→measurement value converter, an electronic processing
unit, a →display, and usually also an →interface (Fig. 46).
The weight force of the sample to be weighed is compen-

sated by a suitable measurement transducer, which converts the weight force into an electrical signal. The resulting output quantity is displayed, and may also be forwarded across the interface.

There are essentially three types of electromechanical weighing instrument (see also →physical weighing principle):

1. Instruments without lever systems with direct electromechanical measurand conversion (e.g. →strain gage or →EMFC load cells).

2. Instruments with lever systems to reduce the weight force that is being measured and for subsequent electromechanical measurand conversion (e.g. →spring load cells (→spring scale), string load cells, →EMFC load cells, or →gyro load cells).

3. Instruments with mechanical weighing-out device and electromechanical measurand conversion (e.g. →deflection balances and →spring scales with measuring wheel, code disc, and electric potentiometer).

The resulting weight force of the load acts either directly or via the lever system on the →load cell as →measurement transducer. The signal of the load cell is a function of the weight force (depending on the principle, generally proportional to the load) and is converted into an analog and/or digital →measurement value, including conversion to mass units and correction of influencing and disturbance variables (e.g. temperature). The data is output on a digital display. In a compact instrument, all of the components are accommodated in a housing. In weighing systems, the individual function blocks can be separated by interfaces, with the data being transmitted via cable or wirelessly. This is often the case when weighments are performed under difficult environmental conditions (industrial applications, product and raw material warehouses, high temperature, radioactively contaminated environment, etc.).

Fig 46
Circuit diagram of an electromechanical weighing instrument with load cell, signal processor, display, and a data interface

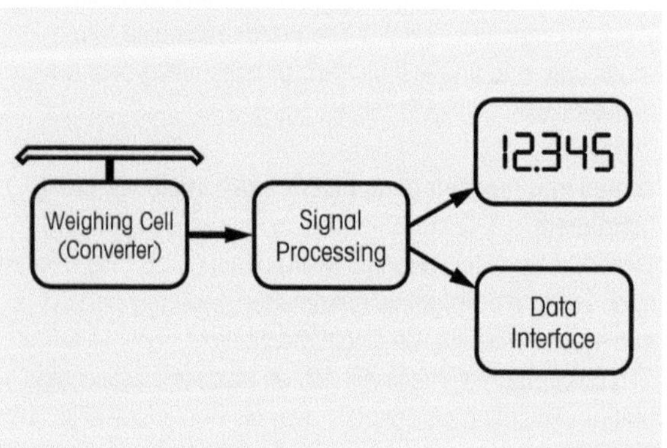

Design Qualification

Part of →Equipment Qualification (EQ). The Design Qualification defines the specifications of the instrument and documents the decision process that results in selection of the supplier and of the instrument.

desorption

The process by which foreign atoms or molecules leave the surface of a solid body. Desorption is the converse of →adsorption and →absorption.

Deutscher Kalibrierdienst

→DKD

deviation

1. Deviation of a value obtained by measurement, and assigned to the →measurand, from the true value. ([DIN 1319-1] 3.5)
2. Value minus its reference value. →measurement deviation
3. Non-technical short form for →systematic error.
4. Non-technical short form for →random error.

dial weight

→Weight piece (in the form of a ring (→ring weight) or compact) that engages on an invariable →lever arm and by means of a setting device that can be actuated from outside; or with the aid of a motor, can be switched in or out. It is used for compensation of the weight force in mass comparison (→physical weighing principle, 1) or as tare preload in →mass comparators. →weight-dialing system

dial weight balance

A →balance in which load equalization is effected wholly or partly by a →weight-dialing system.

dial weight combination

When →weighing with →dial weight balances, highly variable combinations of fractions and multiples of the →unit of mass for the built-in →weight pieces result, depending on the circumstances. The accuracy of the weighing therefore depends heavily on the maximum possible deviation that can occur with each combination.

dialing step

The smallest amount expressed in units of mass that can, for example, be added to or removed from a →dial weight balance.

dialysis scale
→Scale used in dialysis treatment to monitor the weight of the patient while body fluids are exchanged. →bed scale

diet scale
Household →scale, usually taking the form of an →electronic scale (for power-supply or battery operation), used to prepare food for special diets.

differential eccentric load
1. The influence of the →eccentric load on the →net value, i.e. the difference between the eccentric load deviations of the gross value and tare value. For small net weights, the differential eccentric load deviation is considerably less than the eccentric load deviation that applies to the gross value. However, this is on condition that the tare is not removed from the →load receptor, or replaced in the same position, before weighing in.
2. →Specification: The differential eccentric load deviation, i.e. relating to the net value, is usually stated as the standard deviation s_{EC} in mass units, e.g. [g].

differential linearity deviation
→differential nonlinearity

differential nonlinearity
1. The influence of the →nonlinearity on the →net value, i.e. the difference between the linearity deviations of the gross value and tare value. For small net weights, the differential nonlinearity is considerably smaller than the stated maximum nonlinearity, since the linearity deviations of the gross and tare weight are strongly correlated if the latter lie close together.
2. →Specification: The differential nonlinearity, i.e. relating to the net value (→weighed-in quantity), is usually stated as the standard deviation s_{NL} in mass units, e.g. [g].

differential weighing
1. A change (increase or decrease) in mass that is determined by two successive weighments, usually of the same object, to which between the weighments a change was made, and wherever possible with the same balance and same tare weight. →back-weighing
2. Comparison of mass (→mass comparison) of a weighed object with a →mass standard (reference normal).

digit

The smallest indicated →scale interval d on weighing instruments that have a digital display. →readability

digital data processing device

→Electronic device that processes the →measurement signal by digital methods and forwards the weighing results across an →interface in digital form, but does not display them. ([OIML R 76-1] T.2.3.4) (compare: →analog data processing device)

digital device

→Electronic device with digital functions for controlling digital outputs or displays such as a →printer, →auxiliary indicator, keyboard, →terminal, →data memory, or PC. ([OIML R 76-1] T.2.3.4)

digital display

In contrast to an →analog readout, a readout or printout exclusively in the form of numbers, the last digit being rounded. A digital display or digital printout is unambiguously readable and the transfer of measurement values to data processing systems is possible by very simple means, but a digital value cannot be interpolated without additional information and it is difficult to deduce the dynamics of changing values e.g. during →settling.

digital filter

A →signal filter that is implemented with a digital algorithm. A computer uses filter coefficients to form the output value from the current and past input values in real time. Digital filters can suppress low-frequency (~10 Hz) signal components in the weighing signal that are caused by, for instance, air currents, →vibrations in the foundations, or →noise of electronic components. Suppression of these components stabilizes the displayed measurement value. An example of a simple digital filter is one that calculates the arithmetic mean of all incoming values over a specified period of time (→integration time, →measurement time).

digital interval

Difference between two successive digits of equal significance. →readability

digital printout

→digital display

digital-analog converter

An →electronic device for converting digital signals into analog signals (voltages, currents) (Fig. 47). Often referred to as D/A converter. Is used, for example, to plot the progress of a change in weight in analog form with a line plotter. (compare: →analog-digital converter)

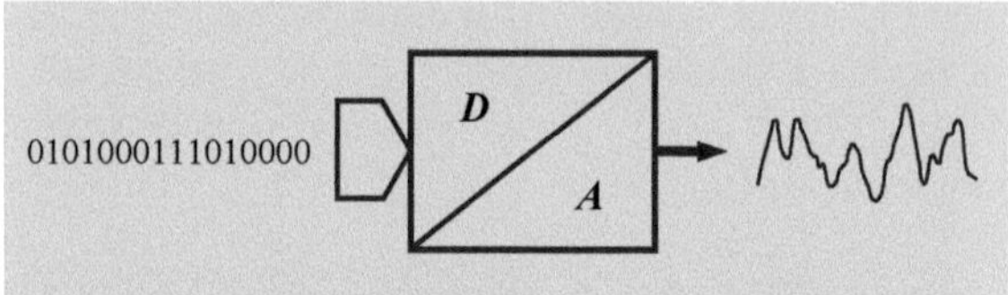

Fig. 47
Digital-analog converter
(left: digital input signal,
right: analog output signal)

Directive on Above-Medium Accuracy Weights

This European Directive stipulates the technical regulations governing the embodiment of →weight pieces of above-medium accuracy with nominal values of 1 g to 50 kg. The permissible →error limits correspond to those of OIML classes E1, E2, F1, F2, and M1. →accuracy classes of weight pieces

Directive on Electromagnetic Compatibility

European Directive for instruments that may cause electromagnetic interference, or whose operation may be impaired by such interference. It stipulates the requirements for protection in this area as well as the corresponding control modalities. The EMC directive is implemented as national law in the EEA and Switzerland. →electromagnetic compatibility, →2004/108/EC

Directive on Machinery

→Machinery Directive

Directive on Measuring Instruments

→Measuring Instruments Directive

Directive on Medium Accuracy Weights

This European Directive stipulates the technical regulations governing the embodiment of →weight pieces of medium accuracy with the following nominal values:
– Bar weights of 5, 10, 20, and 50 kg
– Cylindrical weights of 1, 2, 5, 10, 20, 50, 100, 200, 500 g, and 1, 2, 5, and 10 kg.
The permissible error limits are stated in the directive; only positive deviations from the nominal value are allowed. These error limits vary approximately between those of OIML classes M2 and M3.

Directive on Non-Automatic Weighing Instruments
European Non-automatic Weighing Instruments Directive,
which contains the obligatory and essential requirements
for the harmonized legal and metrological handling of non-
automatic weighing instruments (NAWI). The directive is
implemented as national law in the EEA and Switzerland.
→2009/23/EC

discrimination
Ability of a weighing instrument to react to small variations
of load (not to be confused with →sensitivity). The response
threshold, for a given →load, is the value of the smallest
additional load that, when gently deposited on or removed
from the →load receptor, causes a perceptible change in
the →indication ([OIML R 76-1] T.4.2). The discrimination
is limited by friction and play in the bearings and →joints
of those components of the weighing instrument or load
cell that move relative to each other, by elastic or magnetic
effects, or by the behavior of the electrical measuring system.

dispenser
Device for →volumetric and/or →gravimetric →dispensing,
e.g. vibration dispenser, worm dispenser, manual dispensing balance, worm dispensing balance. →dispensing
balance

dispensing
Separating a quantity into a number of partial quantities
within defined tolerance limits. →dispensing balance,
→fine feed

dispensing balance
→Balance to deliver a preselected quantity of weighing
sample per unit of time by →dispensing a material flow,
usually executed as an →automatic gravimetric filling
instrument. →fine feed, →combination scale

displacement body
Body of known volume used to displace →fluids for the purpose of measuring their density (→density determination).
→plunger, →sinker

display
Element (→module) of an instrument serving to represent
figures, letters, and/or other information (Fig. 48).

display device
→display

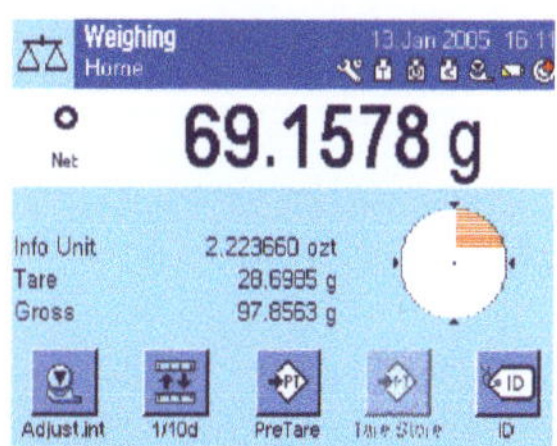

Fig. 48
Display of an analytical balance

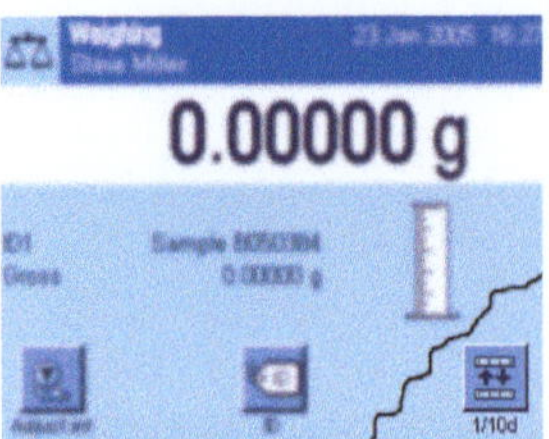
Fig. 49
Balance with switch (bottom right) for changing the readability

display device with reducible resolution
Device which, in response to a manually entered command, temporarily increases the →actual scale interval d, usually by the factor 10 (Fig. 49). →readability, →normal range, →fine range, →extended displaying device

display error
1. →Deviation of the →measurement value.
2. Deviations in the indicated value caused by internal or external interference (e.g. failure of a component).

display screen
→Display on which all forms of character, symbol, and graphics can be represented. May also be connected to weighing instruments, especially when other data in addition to the →weighing results should be displayed.

division
All the divisions of a lined scale or all the numbers of a numeric →scale.

division mark
→Scale mark in the form of lines or dots on the dial.

DKD
Abbreviation for 'Deutscher Kalibrierdienst'. The German Calibration Service is an association of calibration laboratories belonging to industrial companies, research institutes, technical authorities, and supervisory and testing institutions. These laboratories are accredited and monitored by the accreditation office (→accreditation) of the German →calibration service (www.dkd.info).

draft shield
A device that protects the →weighing pan and the →weighed object from disturbing air movements. Draft shields are mainly used on balances with high resolution (→weighing instrument of special accuracy, →weighing instruments of high accuracy). →weighing chamber

drift
The slow change over time in the value of a metrological characteristic (e.g. →indication) of a measuring instrument under constant or stationary conditions. For example, a slowly changing temperature is referred to as →temperature drift.

drift of the measurement value
→drift

dry content
The proportion of solid materials contained in a mixture of solids and liquids, expressed in percent of the total mass of the mixture. →dryer

dryer
Instrument (Fig. 50) for gravimetric (→gravimetry) determination of the →dry content by formation of the difference between the dry weight and the moist weight. The sample is heated by, for example, infrared radiation. The weight lost through drying is measured with a →balance. The heating element can be mounted directly on a →precision balance, or the balance and heating element can be integrated (compact instrument).

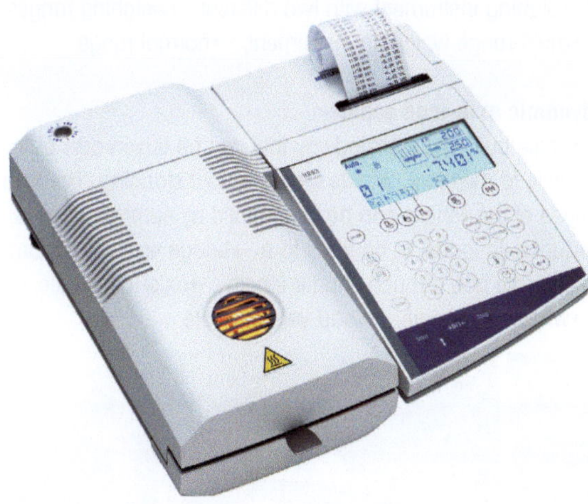

Fig. 50
Dryer with a weighing capacity of 80 g, a readability of up to 0.001% moisture, and a heating range of 40…200 °C

drying oven method
Method for determining the →moisture content of a sample, in which the sample is dried in a drying oven at a constant temperature for a defined period of time. The difference in weight before and after drying is used to calculate the moisture content as a percentage. The drying oven method comprises 13 work steps that include →weighing the weighing glasses with no sample, apportioning (→weighed-in quantity) the sample into the weighing glasses, documentation of the weighing results, and →back-weighing the weighing glasses with the sample after drying, with the corresponding waiting times in-between. The moisture result therefore usually only becomes available after hours or days. (Loss on drying [USP<731>]). →dryer

drying program

Sequence of instructions with which the temperature cycle of a →dryer can be preselected.

Dual Range

Second non-movable range of a weighing instrument that starts at zero and usually has a →readability (→fine range) of only one tenth of the →normal range. →dual range balance, →multi-range weighing instrument (compare: →DeltaRange)

dual-range balance

→Dual-range weighing instrument with a non-movable →fine range. →Dual Range, →dual range balance, →multi-range weighing instrument

dual-range weighing instrument

A weighing instrument with two different →weighing ranges. →multi-range weighing instrument, →normal range

dynamic axle-load scale

→Scale built into the roadway that can be driven over by a vehicle to measure its axle loads without stopping. The total mass of the vehicle can be determined by summing its individual axle loads. In contrast to a →bridge scale, a dynamic axle scale has a much shorter length. →axle-load scale, →wheel-load scale, →road vehicle scale

Fig. 51
Dynamic axle-load scale
(Image by courtesy of Digisens AG, Murten, CH)

dynamic weighing

Weighing of goods that are in motion. The weight force of such goods is augmented by acceleration forces with the result that the total force acting on the weighing instrument changes with time. Example: →hump scale

e

→verification scale interval

e-mark

With this mark, the manufacturer of →prepackages confirms
that the prepackage conforms to the European Packaging
Directive (→Prepackaged Products Directive) (Fig. 52).

Fig. 52
European e-mark for prepackages

EAN

Abbreviation for 'European Article Numbering'. This numbering system allows products to be uniquely identified
internationally by a 13-digit number. In the form of a →bar
code, this number is in commercial use for inventory control.
→UPC

EC Declaration of Conformity (DoC)

With an EC Declaration of Conformity, the manufacturer
confirms that the European Directives applicable to the respective product are fulfilled. These are the →Low Voltage
Directive 2006/95/EC, →EMC Directive 2004/108/EC,
→Directive on Non-Automatic Weighing Instruments
2009/23/EC (where applicable), ATEX Directive →94/9/EC
(where applicable), and →Measuring Instruments Directive
2004/22/EC (where applicable). In addition to issuing an
EC Declaration of Conformity, the manufacturer also affixes
the →CE mark to the instrument. With reference to Directive
→2009/23/EC, the EC Declaration of Conformity confirms
that every weighing instrument that is put into service conforms to the type described in the →EC type approval.

EC Directive

→European Directive(s) …

EC type approval

A certificate issued after testing by a →Notified Body that is
valid in the EEA and Switzerland stating that the construction
of the weighing instrument conforms to the stipulations of
the European →Directive on Non-Automatic Weighing Instruments (→2009/23/EC). The testing procedure, whose successful performance is followed by the issue of the EC type
approval, is described in Directive 2009/23/EC under →EC
Type Examination.

EC type examination

→Type examination applicable in the European Union for
verification of conformity to European Directives that are
relevant to that instrument. Examples: For →non-automatic

weighing instruments according to European Directive →2009/23/EC, the →Notified Body issues an →EC type approval, for →automatic weighing instruments according to European Directive →2004/22/EC the corresponding document is called an EC type examination certificate. For instruments intended for use in hazardous areas (→ATEX 95 Directive), an EC type examination is also a possible means of verifying conformity (→explosion protection).

EC verification

1. Procedure in which a →Notified Body examines each individual →non-automatic weighing instrument for conformity with the type described in the →EC type examination certificate and the basic requirements of the mentioned directive. On determination of the said conformity, the Notified Body issues a declaration of conformity for each individual weighing instrument.
2. Procedure in which the manufacturer or its proxy resident in the EU assures and declares that the →non-automatic weighing instruments that were examined for conformity with the →Directive on Non-Automatic Weighing Instruments fulfill the relevant stipulations of the Directive or correspond to the type described in the respective →EC type approval certificate. The manufacturer or its proxy resident in the EU affixes the →CE mark to each weighing instrument and issues a written →EC Declaration of Conformity. Although this procedure is carried out by the manufacturer, it is designated "EC verification" in practice.

EC verification mark

Synonym for →'Green M'. →EC verification marking, →verification mark

EC verification marking

The marking for →EC verification consists of the →CE mark, and the →Green M to indicate conformity with the →Directive on Non-Automatic Weighing Instruments or the →Measuring Instruments Directive (Fig. 53). →verification mark

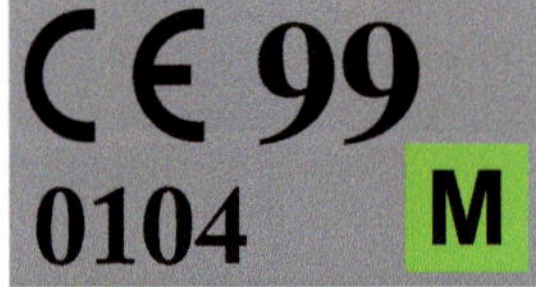

Fig. 53
Marking for EC verification

eccentric load

1. Deviation in the measurement value caused by eccentric loading, in other words asymmetrical placement of center of gravity of the load relative to the →load receptor (Fig. 54). The eccentric load increases with increasing load and distance from the center of the load receptor. →parallel guide

2. →Specification: Magnitude of eccentricity deviation for
    the specified test load and prescribed position (→eccen-
    tric load test), usually expressed as a limit value in mass
    units, e.g. [g]. →differential eccentric load

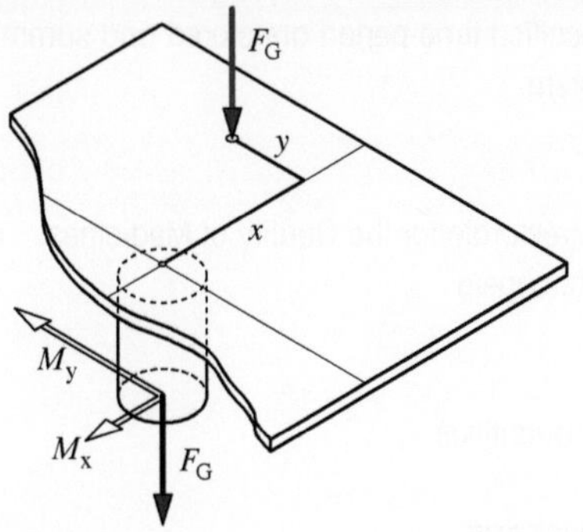

Fig. 54
Eccentric loading causes the load cell
to experience a torque in addition to
the weight force, which may cause a
deviation in the measurement value

F_G: weight force of load
x, y: eccentricity of placement
M_x, M_y: mechanical couples caused
by the eccentric loading

eccentric load deviation
→eccentric load

eccentric load test
Eccentric loading of the →load receptor is performed to determine the extent to which the →measurement value of the
weighing instrument depends on the distribution of the load
on the load receptor. According to [OIML R 76-1] 3.6.2.1
and [NIST HB 44] N.1.3.7., a test load of 1/3 of →nominal
capacity should be placed in the centers of the four quad-
rants (Fig. 55) ([OIML R 76-1] A.4.7.1). →eccentric load

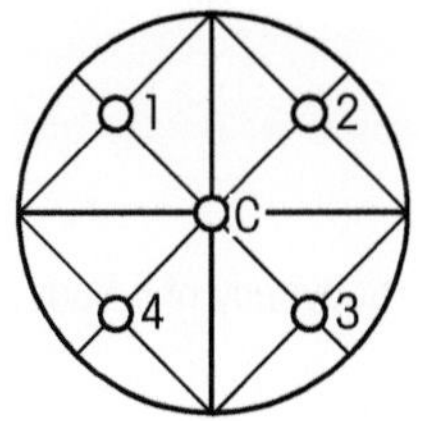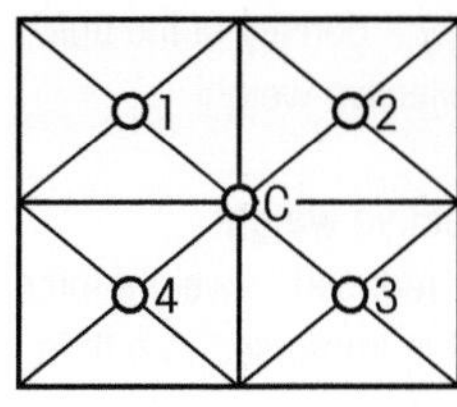

Fig. 55
Positions on the load receptor to test
the eccentric load

C: central loading
1…4: eccentric loadings

eccentric load, differential
→differential eccentric load

eccentric loading
→eccentric load, →load

eccentricity
→eccentric load

eddy-current damping
→damping systems

EDP system

Abbreviation for electronic data processing system. Program controlled computer on which large volumes of data are stored and/or processed. For example, all weighing results of weighing instruments and auxiliary devices that occur during a specified time period are stored and summarized in an EDP system.

EDQM

European Directorate for the Quality of Medicines. →European Pharmacopeia

effect

→influence quantities

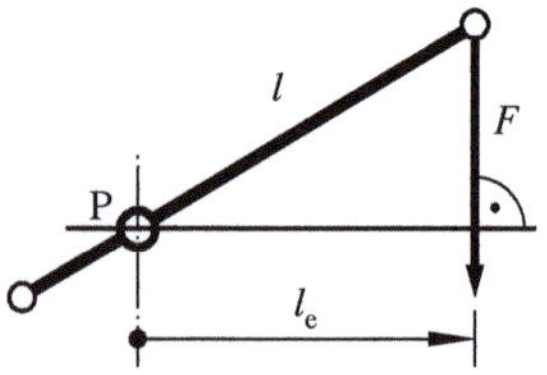

effective lever arm

Perpendicular distance l_e between the line of action of a force F acting on a lever and the fulcrum P of the lever (Fig. 56).

effective mass

→Buoyancy causes the mass of a body that is immersed in a fluid to appear to be less than when it is determined from the →weight force

$$m^* = m - V\rho_F$$

m^* effective mass of the body
m mass of the body
V volume of the body
ρ_F density of the fluid
→effective weight

effective weight

The reduced →weight force due to →buoyancy of a body that is immersed in a fluid

$$G^* = mg - V\rho_F g = (m - V\rho_F)g = m^* g$$

G^* effective weight of the body
g gravity
m mass of the body
m^* effective mass of the body
V volume of the body
ρ_F density of the fluid
→effective mass

electric charge

Physical quantity which, among other things, is the source of the electric field and of force effects (Fig. 57). The electric

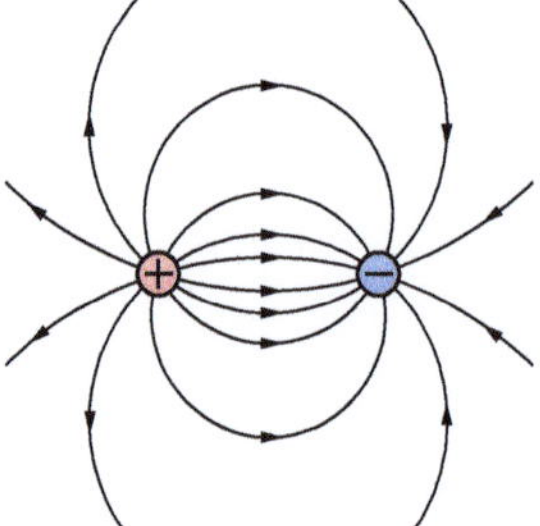

Fig. 56
Effective lever arm

F: force
l: lever arm
l_e: effective lever arm
P: fulcrum

Fig. 57
Electric field of two point charges of different polarity

charge can be positive or negative (polarity); charges with
opposite polarity attract, those with the same polarity repel.

electrical safety
Non-technical term for the requirements of the →Low Voltage
Directive.

electrodynamic converter
Physical →converter principle in which an electric conductor
is located in a magnetic field (Fig. 58). When the current i
flows through the conductor, a force F acts on it

$$F = c{\cdot}i.$$

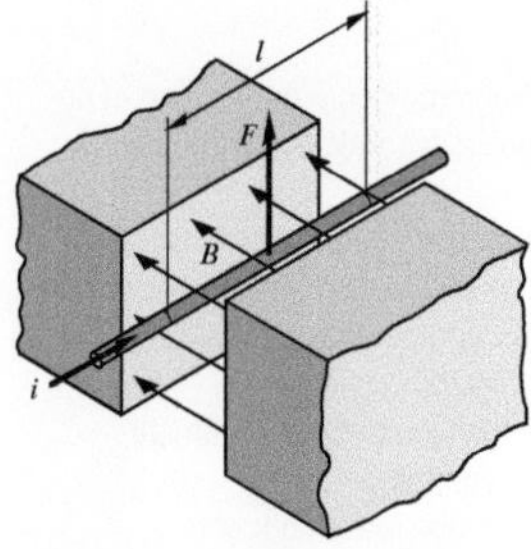

Fig. 58
Electrodynamic converter principle

B: magnetic flux density in the air gap
i: electrical current in the conductor
l: length of conductor in air gap
F: electrodynamic force

The proportionality factor c (a.k.a. conversion factor)
depends on the magnetic flux density B, the length l of
the conductor and its orientation relative to the magnetic
field. Provided the conductor and the magnetic field are
orthogonal to each other, the proportionality factor amounts
to

$$c = Bl.$$

The direction of the force is always orthogonal to both the
conductor and the field. The electrodynamic converter has
many different applications, for example, electric motors and
generators, loudspeakers, sensors, measurement converters,
→EMFC load cells, etc.

electromagnetic compatibility (EMC)
The ability of an electrical apparatus, device, or system to
function satisfactorily (i.e., comply with the →error limits)
within its electromagnetic environment without itself causing
electromagnetic interference that would be unacceptable for
all apparatuses, equipment, or systems in this environment.
In the European Union, the requirements for instruments are
regulated by the →European Directive on Electromagnetic
Compatibility →2004/108/EC.

electromagnetic force compensation
Weighing principle (→physical weighing principle) in which
the →weight force of the weighed object is opposed by a
force of equal magnitude (→force compensation) that is
produced with the aid of an →electrodynamic converter
(Fig. 59 and 60). An electronic control system that responds
to the displacement of the load cell (position sensor) causes
the compensation current to be adjusted so that a disequi-
librium of the forces arising from loading or unloading is
restored to equilibrium. The weight force is proportional to
the mass of the weighed object and in the stable state is

completely compensated by the electrodynamically generated force. This force is itself proportional to the electric current flowing in the converter. The current is thus proportional to the mass placed on the balance and can therefore by used as a →measurement signal.

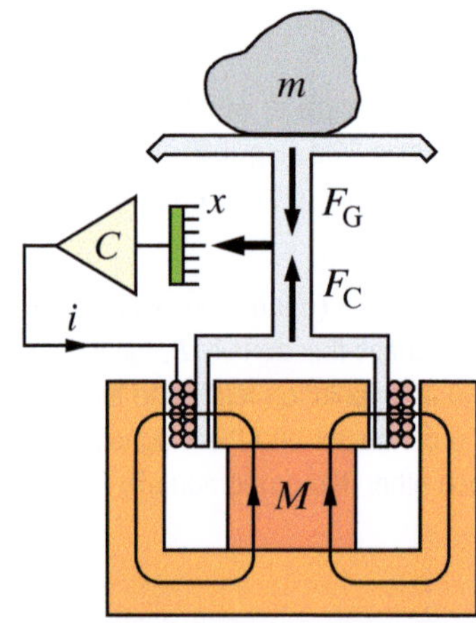

Fig. 59
Diagrammatic cross section of an electrodynamic compensator:

m: mass of the weighed object
F_G: weight force of the weighed object
F_C: compensation force of the electrodynamic converter
x: (temporary) displacement
C: controller
i: compensation current
M: permanent magnet

Fig. 60
Cutaway view of the electrodynamic force compensator of an analytical balance
The position sensor can be seen at the top. In the middle, in the air gap of the magnetic circuit, is the compensation coil that creates the compensation force. Visible in the lower part of the picture is the permanent magnet that creates the necessary magnetic flux in the air gap, which flows through the magnetically conducting (iron) parts back to the magnet.

electromechanical weighing instrument

A weighing instrument that uses electromechanical means of load compensation. The weight of the load as →measurand is compensated with an electromechanical converter and evaluated by electronic means (e.g. →strain gage scale, →EMFC weighing instrument, →string balance). →design and function of an electromechanical weighing instrument

electronic assembly

Part of a device that contains electronic components and has an identifiable independent function, e.g. →analog-digital converter, →digital display.

electronic device

Device with →electronic assemblies on or in the weighing instrument that performs a special function, or represents the instrument itself ([OIML R 76-1] T.2.3.1).

electronic weighing instrument

→electromechanical weighing instrument

electrostatic charging

→Electric charge that many accumulate on objects (solids, liquids, or gases) that have low electrical conductivity (so-called electric insulators) such as glass, plastic, organic solvents, seeds, powders, dusts, etc.

electrostatic discharging

When electrostatically charged people or objects touch parts of a weighing instrument, an instantaneous electrostatic discharge can occur (discharge voltage several kilovolts, energy several millijoules). This causes a momentary discharge current to flow through the instrument that has an order of magnitude of several kiloamperes and which can adversely affect the correctness of the measurement value determination or even destroy electronic circuits.

electrostatic influence

When weighing instruments are being used, electrostatic forces resulting from →electrostatic charging of the weighing sample, or less frequently also of parts of the instrument (load pan, draft shield, housing, etc.) can influence the weighing. Since all vertical forces acting on the weighing pan, irrespective of their source, are interpreted by the instrument as mass, the electrostatic force can cause substantial errors in indicated values depending on the shape and size of the item being weighed (0.1 g on an analytical balance is not uncommon) (Fig. 61). Depending on the conductivity of the weighing sample, the time required for the →measurement value to drift towards the correct weighing value is shorter or longer (seconds to hours). →drift

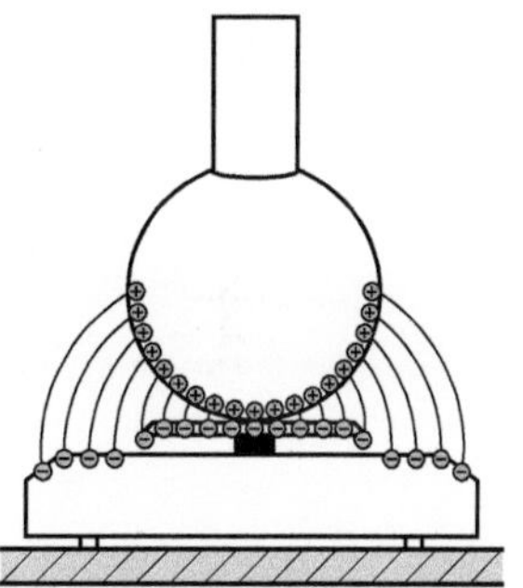

Fig. 61
Electrostatic influence on a vessel being weighed

EMC

Abbreviation for →'electromagnetic compatibility'.

EMC Directive

→Directive on Electromagnetic Compatibility

EMFC

Abbreviation for →'electromagnetic force compensation'.

EMFC load cell

Load cell that uses →electromagnetic force compensation as measurement principle. EMFC load cells are mainly used in →high-resolution balances such as →analytical balances and high-grade →precision balances. →design and function of an electrodynamic balance

EMFC weighing instrument

An →electromechanical weighing instrument in which an →EMFC load cell is used as measurement →converter. →physical weighing principle

EMFR

Abbreviation for 'electromagnetic force restoration'. →electromagnetic force compensation

EN 45501

→European Standard EN 45501

EN 60529

European Standard EN 60529, →degrees of protection provided by enclosures

endurance of the printout

Certification requirements exist for the printout of weighing results, e.g. printouts for the intended purpose must be clear and enduring, i.e. good readability must be assured for a minimum of
a) two years under the usual archiving obligations;
b) one month for price-mark devices and industrial scales;
c) one week for scales for →public points of sale even under adverse conditions (contact with grease or food, effect of light).

engineering standards

Whereas the legal certification requirements are specified in laws, statutory regulations, and directives, in the requirements for the design of measuring instruments, reference is made to the recognized engineering standards. Recognized engineering standards are, for example, standards, test rules, or recommendations of →Notified Bodies or →national metrology institutes.

environmental influence

External circumstances that may adversely affect the metrological behavior of a weighing instrument. These may include, for example, setting up, →ambient temperature, air current, weather conditions, magnetic fields, electrostatic forces, and →vibrations. →influence quantities

equal-arm beam balance
→Beam balance, the main →bearing of which is in the middle of the →balance beam and whose levers are thus of equal length (Fig. 62 and 121). Today, in technological fields, this type of balance is used only for special applications, but historically it is the icon of weighing technology as well as a symbol of justice (Fig. 63).

equilibration of the weighing instrument
→settling

equilibrium
1. Equilibrium state of a system. →equilibrium position
2. State of two forces that are equal and opposite, and therefore mutually neutralizing. →force compensation

equilibrium position
1. That position of a rotatable, rigid body in the gravitational field in which it is not subject to any torque caused by →gravity (Fig. 64).
1.1 Stable equilibrium: The center of gravity of the body is situated vertically below the fulcrum (e.g. pendulum).
1.2 Unstable equilibrium: The center of gravity of the body is situated vertically above the fulcrum (e.g. tightrope walker).
1.3 Indifferent or neutral equilibrium: Center of gravity and fulcrum of the body coincide (e.g. wheel).

Fig. 62
Equal-arm beam balance

Fig. 63
Equal-arm balance as symbol of justice
(Justitia statue in Frankfurt/Main, DE)

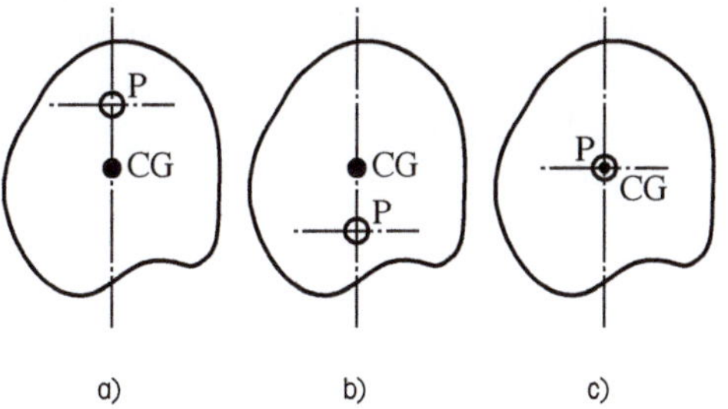

Fig. 64
Equilibrium positions
a) stable;
b) unstable;
c) indifferent or neutral equilibrium

CG: center of gravity
P: pivot point

Equipment Qualification
Officially undefined but commonly used term in instrument qualification. Qualification is the process that proves and documents that an instrument functions correctly and delivers the expected results (EU Guide on Manufacture). Equipment Qualification (EQ) contains the following qualification steps: →Design Qualification (DQ), →Installation

Qualification (IQ), →Operational Qualification (OQ), →Performance Qualification (PQ), and →Maintenance Qualification (MQ).

equivalence principle
The equivalence principle states that in a gravitational field (→gravitation) in a vacuum, all bodies, irrespective of their properties such as →mass, shape, or material, describe the same path (→acceleration due to gravity). Gravitational mass and inertial mass are thus equivalent.

error
→measurement deviation

error due to the display
1. →display error
2. Obsolete term for →measurement deviation.

error limit class
Classification of →measuring instruments or →standards into classes according to specified error limits, e.g. weight pieces according to OIML Recommendation R 111. →OIML weight classes.

error limit component
The error limit component p_i defines the applicable error limits for separately inspected →modules of a weighing instrument. The error limits of a module are identical to the component p_i of the error limit of the entire weighing instrument. The error limits of a module must be related to the same accuracy class and number of verification scale intervals as apply for the complete weighing instrument.
The components p_i of the individual modules of a weighing instrument must satisfy the equation

$$p_1^2 + p_2^2 + p_3^2 + \ldots \le 1$$

([OIML R 76-1] 3.10.2.1).

error limits
Maximum amounts for positive or negative deviations. Error limits are mainly specified in relation to →systematic errors of the measurement values from the correct value or from another specified or agreed value of the measurement value. It is not permitted for error limits to be exceeded by →random errors. There is a need to differentiate between, for example, →maximum permissible error on verification and →maximum permissible error in service. →maximum permissible error

error, random
→random error

error, systematic
→systematic error

European Declaration of Conformity
→EC Declaration of Conformity

European Directive Concerning Equipment and Protective Systems Intended for Use in Potentially Explosive Atmospheres
→ATEX 95 Directive, →94/9/EC

European Directive on Electromagnetic Compatibility
→Directive on Electromagnetic Compatibility, →2004/108/EC, (→89/336/EEC)

European Directive on Good Laboratory Practice
→Good Laboratory Practice, →2004/9/EC, →2004/10/EC

European Directive on Good Manufacturing Practice
→Good Manufacturing Practice, →2003/94/EC

European Directive on Measuring Instruments
→Measuring Instruments Directive, →2004/22/EC, →automatic weighing instrument

European Directive on Non-Automatic Weighing Instruments
→Directive on Non-Automatic Weighing Instruments, →2009/23/EC, (→90/384/EEC)

European Directive on Prepackaged Products
→Prepackaged Products Directive, →76/211/EEC

European Directive on requirements for safety and health protection of workers at risk from explosive atmospheres
→ATEX 137 Directive, →1999/92/EC

European Directive Relating to Electrical Equipment Designed for Use Within Certain Voltage Limits
→Low Voltage Directive, →2006/95/EC, (→73/23/EEC)

European Directive relating to medium accuracy weights
Rectangular →bar weight and →cylindrical weight directive.
→Directive on Medium Accuracy Weights, →71/317/EEC

European Directive relating to weights from 1 mg to 50 kg of above-medium accuracy
Above-Medium Accuracy Weights Directive. →Directive on Above-Medium Accuracy Weights, →74/148/EEC

European Machinery Directive
→Machinery Directive, →2006/42/EC (→98/37/EC)

European Pharmacopeia
Directory of medicines that was given legal status by the signatory states of the European Pharmacopoeia Convention. The European →Pharmacopoeia is published, updated, and expanded by the European Directorate for the Quality of Medicines (EDQM). The mission of the European Pharmacopoeia Convention is to harmonize the standards of the various national European Pharmacopeias as well as the quality standards and control processes for medicines (www.pheur.org).

European Standard EN 45501
European Standard EN 45501 "Metrological aspects of non-automatic weighing instruments" adds detail to the →Directive on Non-Automatic Weighing Instruments. The directive itself only covers the harmonization of legally binding metrological and technical requirements for →non-automatic weighing instruments that are used in applications subject to legal metrology. EN 45501, which is based on →OIML Recommendation R 76-1, contains metrological as well as design and constructional stipulations for non-automatic weighing instruments, whose fulfillment is indicative of compliance with the main requirements of the above-mentioned directive.

evaluation device
Device in which analog-digital conversion (→analog-digital converter) of the output signals of one or more →load cells (e.g. →strain gage load cell, →string load cell, or electromagnetic force compensation load cell, →EMFC load cell) and the further processing of the data into the weighing result are performed in such manner that display devices and auxiliary devices can be controlled with signals that correspond to the mass (weight).

evaporation
Transformation of a substance from a liquid to a gaseous state without it being heated to boiling point. An evaporating substance (e.g. water, alcohol) constantly loses mass. Because of this, and depending on the →resolution of the

weighing instrument, when such substances are weighed, no stable →measurement value can be expected. See →hygroscopic weighing sample.

exceptions to compulsory verification
The legislative authorities have exempted certain weighing instruments and auxiliary devices from the obligation to being verified, e.g. scales in agricultural operations with →*Max* up to 3 t, →counting scales or coin roll scales. Further exceptions are non-interacting →auxiliary devices if the scale has, for example, an →alibi printer or certified →data memory. This regulation applies in the EEA and Switzerland. →Directive on Non-Automatic Weighing Instruments

expansion factor
Factor k that expands the →standard uncertainty u into the →uncertainty interval U:

$$U = k \cdot u$$

→measurement uncertainty

explosion protection
Measures to avoid hazardous explosive mixtures of gas and air, or dust and air, and to avoid effective ignition sources. In Europe, the intended use of equipment and protective systems in hazardous areas is legally regulated for manufacturers by the →ATEX 95 Directive and for operators by the →ATEX 137 Directive. The manufacturer must fulfill the essential health and safety requirements defined in the ATEX 95 Directive by suitable construction of the equipment or protective system, and demonstrate fulfillment by corresponding tests. The operator is responsible for avoiding explosive atmospheres and for using suitable equipment or protective systems as intended at the place of installation. Such equipment and protective systems may only be used in hazardous areas if they are constructed in such manner that they cannot act as effective ignition sources.

Fig. 65
Official symbol for marking equipment and protection systems intended for use in hazardous areas

1. Hazardous areas:
   At European level (ATEX 137 Directive) and international level (IEC EN 60079-10 for gas and IEC EN 61241-10 for dust), hazardous areas are classified according to their probability of occurrence as follows:
   zone 0 (zone 20): An area where an explosive atmosphere (gas/air, dust/air mixture) is present continuously or for long periods or frequently.
   zone 1 (zone 21): An area where an explosive atmosphere (gas/air, dust/air mixture) is likely to occur in normal operation occasionally.

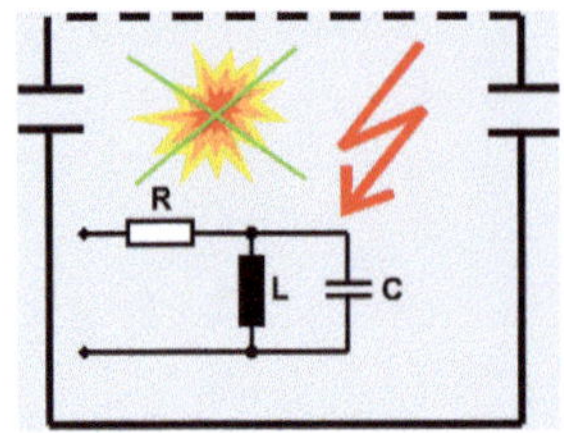

EEx i)

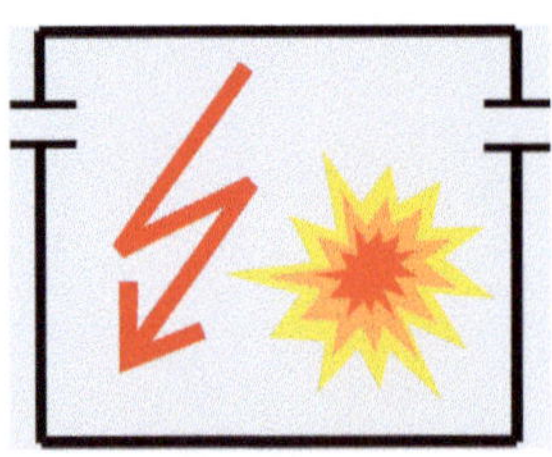

EEx d)

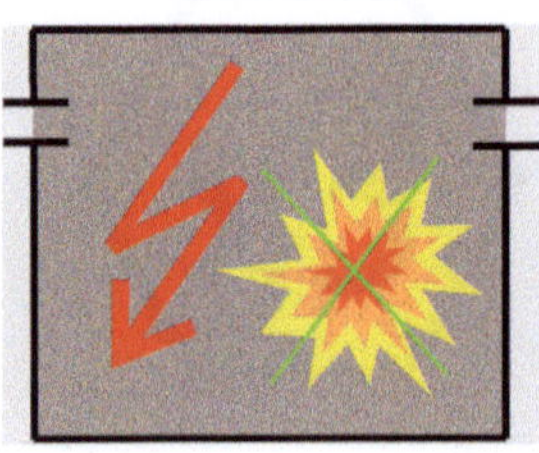

EEx q)

EEx p)

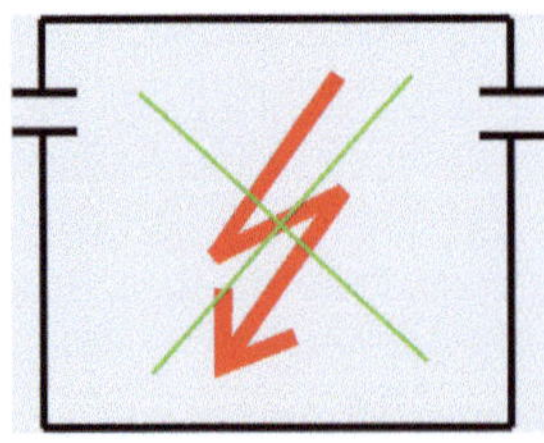

EEx e)

zone 2 (zone 22): An area where an explosive atmosphere (gas/air, dust/air mixture) is not likely to occur in normal operation or will persist for a short time only.

2. Ignition sources:

According to the IEC EN 60079 series "Electrical Installations for Hazardous Areas", the types of protection against ignition of gas explosions are classified as follows (Fig. 66):

EEx i: intrinsic safety

EEx ia: intrinsic safety for the occurrence of two independent faults (for zone 0)

EEx ib: intrinsic safety for the occurrence of one fault (for zone 1)

EEx d: flame proof encapsulation (for zone 1)

EEx q: powder filling (for zone 1)

EEx p: pressurized enclosure (for zone 1)

EEx e: increased safety (for zone 1)

EEx o: oil immersion (for zone 1)

EEx m: encapsulation (for zone 1)

EEx n: ignition protection n (for zone 2).

For protection against dust explosions, (IEC/EN 61241 Series "Electrical Equipment for Use in Presence of Combustible Dust"), in many cases the housing is designed and constructed in such a manner that no dust can penetrate into the instrument and thereby cause an explosion ("Protection by enclosures"). This is achieved by a corresponding →IP protection and other measures. Also in Europe, there are constructional stipulations for non-electrical equipment (EN 13463 Series "Non-electrical Equipment for Potentially Explosive Atmospheres).

At European level, the ATEX 95 Directive divides equipment into three separate categories (category 1 with very high protection for use in zone 0 (zone 20), category 2 with high protection for use in zone 1 (zone 21), category 3 with normal protection for use in zone 2 (zone 22). For example, equipment with protection type "EEx ia" that is used in zone 0 is assigned to category 1G (G for gas, D for dust).

The following types of protection are typically used on weighing instruments:

"intrinsic safety EEx ib" for zone 1 (ATEX 95: category 2G)
"energy limitation EEx nL" or "non-sparking EEx nA" for zone 2 (ATEX 95: category 3G)
"Protection by enclosures" for zone 21 and 22
(ATEX 95: Categories 2D and 3D)

extended displaying device

A device that temporarily changes the →actual scale interval d to a value less than the →verification scale interval e in response to a manual command ([OIML R 76-1] T.2.6).
→readability, →display device with reducible resolution

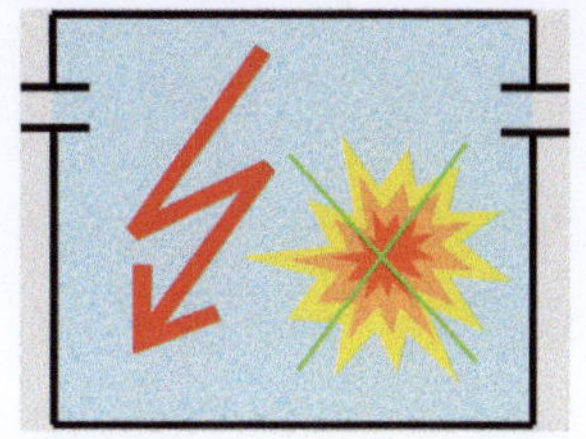

EEx o)

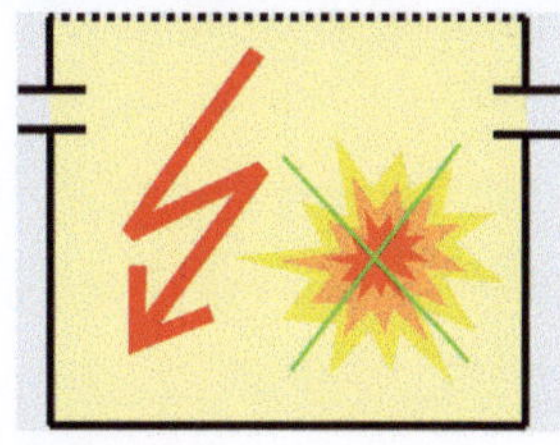

EEx m)

Fig. 66
Types of protection
EEx i) intrinsic safety;
EEx d) flameproof encapsulation;
EEx q) powder filling;
EEx p) pressurized enclosure;
EEx e) increased safety;
EEx o) oil immersion;
EEx m) encapsulation

FACT

Acronym for 'Fully Automatic Calibration Technology' (vendor-specific name). →Automatic adjustment of the →sensitivity, sometimes also the →linearity, of a weighing instrument. Adjustment is triggered after expiration of a defined time period from switch-on and/or on exceeding a defined temperature change (Fig. 67). →Autocal, →proFACT

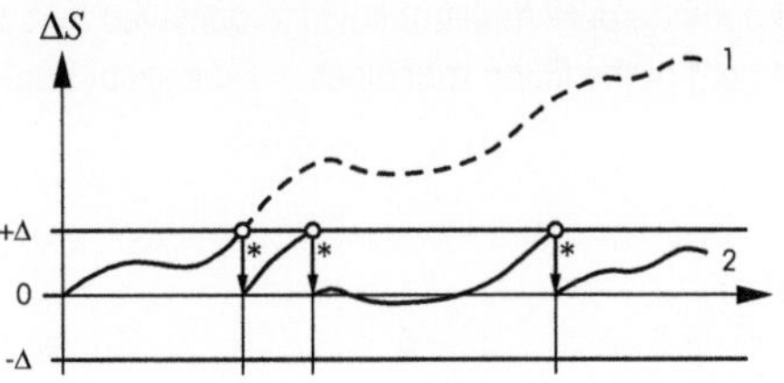

Fig. 67
Fully-automatic sensitivity adjustment. The chart shows a possible progression of the sensitivity deviation over time (1) without, and (2) with, automatic adjustment of the sensitivity.

t: time (since last adjustment)
ΔS: sensitivity offset
Δ: magnitude of largest offset
*: instant of adjustment

family

A group of →weighing instruments or →modules of the same type series with the same measurement principle and constructional characteristics (e.g. identical control unit, identical construction of load cells and force transmission elements, but different metrological data, such as →*Min*, →*Max*, →*e*, →*d*, →accuracy class). ([OIML R 76-1] T.3.5)

fill quantity

A term used in →prepackage process control:
1. →Nominal fill quantity to designate the amount of material indicated on the →package.
2. →Target fill quantity, which designates the average fill quantity that must be maintained to ensure that the legal regulations are not violated.

filling process control

Inspection of the →fill quantity of →prepackages performed by →Weights and Measures authorities. The requirements regarding the fill quantity are defined in the European Packaging Directive (in Germany, in the →Weights and Measures Act and →Prepackaged Products Decree). →filling process control facility

filling process control facility (FPC)

A partially or fully automated facility designed for the filling process control of →prepackages (Fig. 68). The system may consist of a →precision balance and a freely programmable computer, or a precision balance with a calculating printer. In larger facilities, several weighing instruments are connected to a process computer, possibly over relatively large distances. In addition, fully automated installations are equipped with automatic package conveyors. In such cases, the precision balance can be replaced by a →check-weigher. With systems of this kind, the requirements of the →Prepackaged Products Directive can be satisfied quickly and objectively and the necessary documentation provided without delay. At the same time, the filling stations are moni-tored, and considerable material savings achieved by cor-rect adjustment of the filling machines. →statistical quality control

Fig. 68
Checkweigher for prepackages

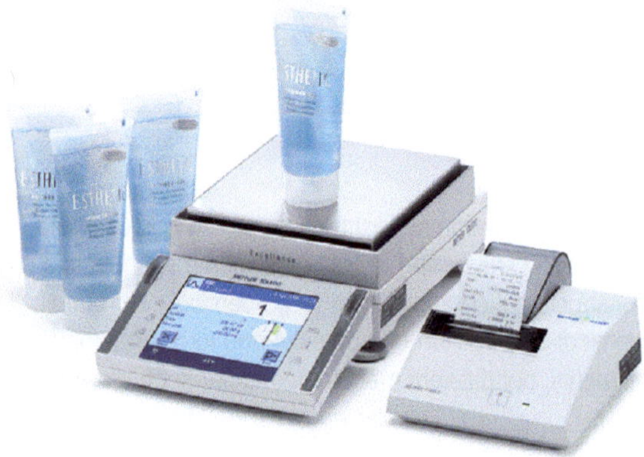

filling scale

An →automatic gravimetric filling instrument when used for filling operations. The weighing sample is automatically fed to the scale, and then usually weighed into batches of equal amounts, and usually also automatically conveyed to the next station for final packaging. →prepackage process con-trol, →combination scale

filter

1. Device for separating different media, e.g. plastic or glass fiber that is used to trap particles contained in the air (→filter balance).
2. Electronic device for separating signal components of dif-ferent frequency. The characteristics of the filter are deter-mined by filter parameters that define the frequency and phase patterns. In weighing instruments, filters are used to suppress interference that may be present in the signal of the weighing sensor.

filter balance

Balance suitable for weighing particle →filters (Fig. 69). The filter is weighed before and after being used, so without and with the particles that are trapped in the filter. From the dif-ference, the quantity of substance trapped in the filter can be

determined. Since the quantity is usually very small, either
→semimicro balances or →microbalances are generally
used. Great care must be taken in this application to avoid
→electrostatic influence, since most particle filters are poor
electric conductors and can therefore become electrostatically charged.

final weight value
The weight value that is obtained when the weighing instrument is completely at rest and balanced, with no disturbanc-
es affecting the indication ([OIML R 76-1] T.4.6). →settling,
→stability

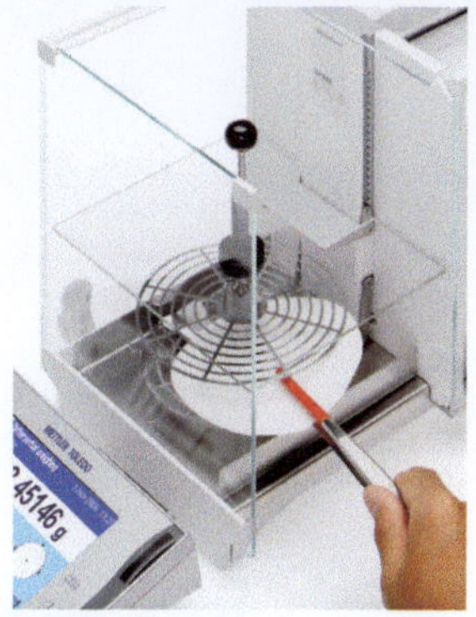

Fig. 69
Filter balance

fine adjuster
A manually adjustable device that serves to subdivide the
scale intervals of a self-equilibrating or →semi-self-indicat-
ing instrument that is equipped with a movable line scale.
After the instrument has equilibrated, the exact distance
between a division line and the zero mark is determined
by means of a step-up device that can enlarge the reading
mechanically, optically, or electrically.

fine dispensing
→Gravimetric or →volumetric distribution of a constant or
discrete material flow. Coarse dispensing is followed by
fine dispensing to ensure that the prescribed distribution is
achieved as accurately as possible. →fine feed

fine feed
The material to be weighed that is fed to an automatic
weighing instrument for apportioning (→apportion) into the
→load receptor container with which the set fill quantity is
attained at the end of the filling or dispensing operation. The
fine feed is preceded by the →coarse feed. →fine dispensing

fine range
1. Fine range: Range with lower readability (additional display position or positions) for the output of the measuring result than in the →normal range (the latter is often
   referred to as the →coarse range). →DeltaRange, →Dual
   Range
2. For verified weighing instruments with an →auxiliary
   indicating device, a distinction is made between the
   →verification scale interval e and the →actual scale
   interval d. In practice, the auxiliary indicating device is
   often described as fine range. The additionally displayed
   digit(s) are marked, e.g. hatched.

fine weight

Term defined in the German →Verification Ordinance for →weight pieces of class OIML F1 (→OIML weight classes) ([VO] Appendix 8, Section 2, 2.1).

firmware

Program (→software) for the dedicated control of an instrument that is stored in permanent memory. Depending on the memory technology, the firmware may be capable of being updated.

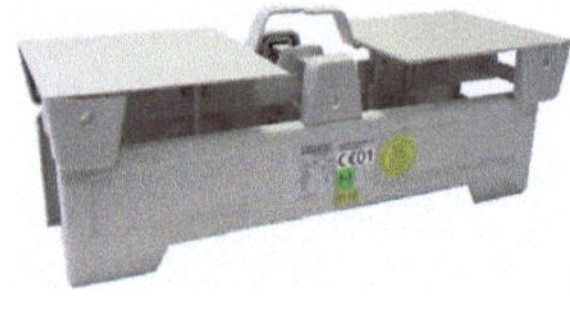

Fig. 70
Flat-pan scale
(Image by courtesy of RHEWA,
August Freudewald GmbH & Co. KG,
Mettmann, DE)

flat-pan scale

Equal arm (→equal-arm beam balance), →top-loading →bridge scale, usually executed as a →bench scale with a →weighing capacity of up to 20 kg, whose →load receptor is in the form of a flat pan (Fig. 70). The →Béranger scale, →Pfanzeder scale, and →Roberval scale are examples of flat-pan scales.

flexible bearing

→flexible joint

flexible coupling

Monolithic, elastic coupling (→joint) that guides mechanical parts in relation to each other. Limited movement between the guided parts occurs through elastic deformation of the coupling. →flexible joint, →pivot joint, →cross-flexed spring joint

flexible joint

→Flexible coupling between two parts that move relative to each other. By means of elastic deformation, the joint allows a limited amount of tipping (flexure) of the coupled parts about the axis of rotation perpendicular to the longitudinal axis (flexible spring). Under tensile forces, and to a lesser extent also under compressive forces, the joint behaves practically as a rigid coupling. Flexible joints are usually made of a special alloy or, in the case of monolithic load cells (→Monobloc), from the same material as the other components of the load cell. To create a defined flexure point, a particular shape must be given, which is usually produced by machining, cold forming, or spark erosion. Flexible joints are maintenance-free and characterized by low internal friction and robustness; they are a prerequisite for high repeatability of a load cell. One version of a flexible joint is the →cross-flexed spring joint.

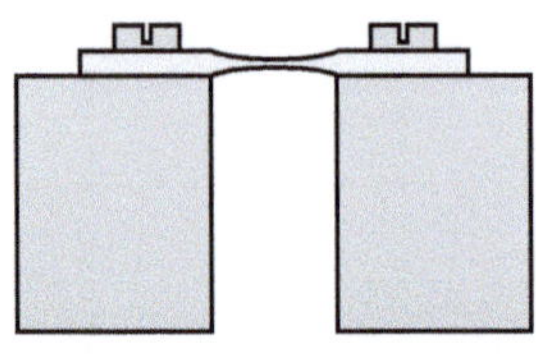

a)

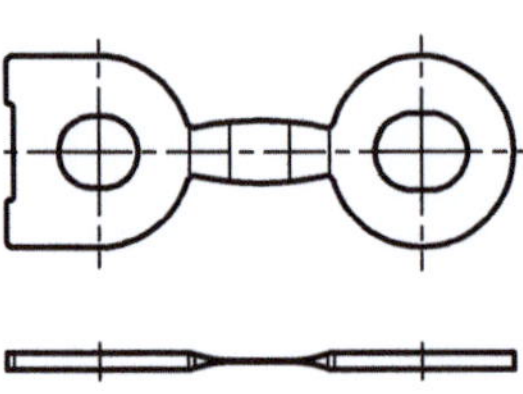

b)

Fig. 71
Flexible joint
a) operating principle;
b) exemplary embodiment

flexure pivot
→flexible joint

floor scale
Freestanding →bridge scale for installation on the floor.

fluid
Substance possessing the physical property of being able to
flow. Gases and liquids are fluids.

foot switch
An operating element used to →tare the weighing instrument
or trigger a data transfer. It allows the operator to use both
hands for other tasks.

force
Term first used by Archimedes for the physical quantity that
is the cause of all motion or change in shape. There are
many sources of force including deformation (→spring
force), →gravitation (→weight force, →buoyancy), electrostatic and dynamic forces, magnetic forces (→magnetism),
kinetic forces (acceleration force, centrifugal force), and friction. The SI unit of force is the →newton.

force comparison
Determination of the →weight force of the load by means
of another →force that is not a weight force, as for example
a deformation force (→spring scale) or electromagnetic
force (→electromagnetic force compensation). In contrast
to →mass comparison, in this →measurement principle the
result depends on →local gravity. →physical weighing principle, →force compensation

force compensation
When weighing, the →weight force of the load is held in
equilibrium by a →force of equal magnitude that acts in the
opposite direction. This compensating force can take various
different forms, e.g. weight force of weight pieces, electromagnetic force (→electromagnetic force compensation), etc.
→physical weighing principle

force due to gravity
Synonym for →'weight force'.

force link
The connecting link that serves to transmit forces between a
→parallel guide and a lever, or between two levers
(e.g. pressure link or tension link) without any connection

to the →frame. In many cases force links take the form of elastic →pivot joints, while pressure links take the form of a →knife-edge and a →pan.

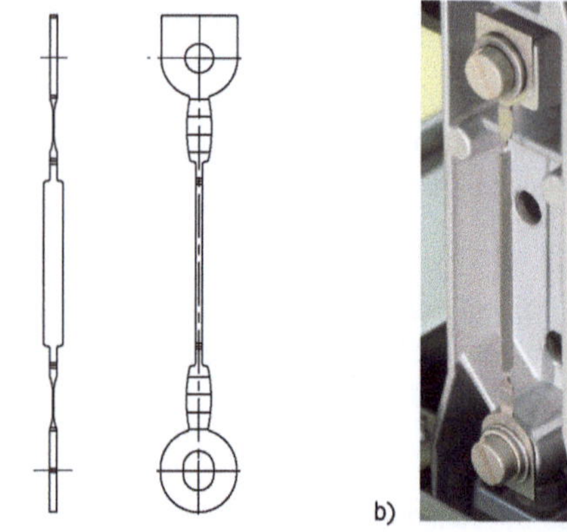

Fig. 72
Tension force link
a) diagram of a tension force link;
b) tension force link used for transfer of force from the hanger (bottom) to the lever (top) of a weighing instrument

force measuring cell
→Measurement transducer that converts the input quantity →force into, for instance, an electrical output quantity.
→load cell

forklift scale
→Scale that is built into a forklift truck for weighing items that are being transported on pallets (Fig. 73). →pallet scale

form printer
→printing device

formula weighing
Weighing of different components into a container with taring (→tare) being performed before each new component is weighed in.

formula weighing system
One or more weighing instruments with special functions for the preparation of formulations, usually with a PC or EDP system for specification of the formulation steps.

FPC
Abbreviation for →'filling process control facility'.

frame
That part of a weighing instrument that carries the support bearings and connects them to each other, as well as to the weighing-out device which usually stands on, or is fastened to, a stable surface (e.g. a bench or the floor respectively).

Fig. 73
Forklift scale
(Image by courtesy of Dini Argeo, Spezzano di Fiorano, Modena, IT)

G

Symbol for →gross value. →B

gage factor

Ratio of the relative resistance change $\Delta R/R_0$ of a →strain gage to the applied strain ε

$$ GF = \frac{\dfrac{\Delta R}{R_0}}{\varepsilon} $$

GF gage factor
ΔR resistance change caused by strain
R_0 resistance of relaxed gage
ε strain applied to gage

For metallic foil gages, the gage factor is usually around 2. A simplified explanation for this is that stretching a wire not only increases its length, but also decreases its cross section by about the same factor (Fig. 158b). These two changes in combination increase the resistance by approximately a factor of two.

galvanic separation

Separation between two electrical units, e.g. a data source and a data sink, or an electricity supply and an appliance, so that electric current cannot flow between them but data and/or energy can. Information can, for example, be transmitted by means of an optocoupler while electric power is transmitted through a transformer. Galvanic separation is used to prevent electrical interaction between peripheral devices and measuring instruments.

gamma sphere

Spherical →plunger, usually made of metal, used to determine the →density of liquids or pasty substances (Fig. 74).

GAMP

Abbreviation for →Good Automated Manufacturing Practice.

garbage scale

→Scale used to determine the weight of garbage and the tariff for its disposal (Fig. 75), usually taking the form of a →skip scale or →vehicle on-board weighing system.

gauge factor

→gage factor

Gaussian distribution

→normal distribution

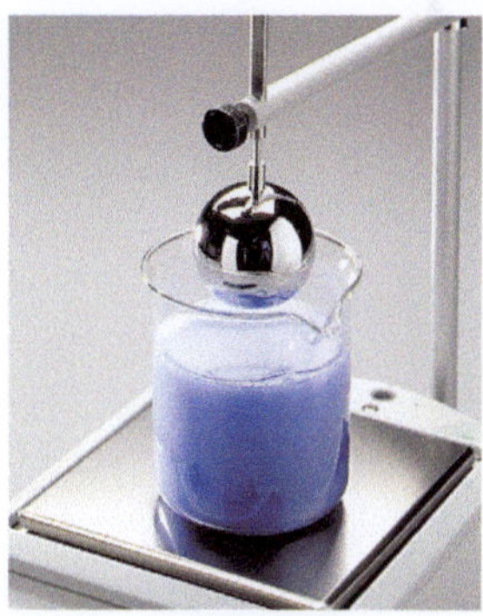

Fig. 74
Gamma sphere

Fig. 75
Garbage truck with integrated scale
(Image by courtesy of Digisens AG, Murten, CH)

Gaussian weighing method

A weighing method in which the sample (unknown load) on the →load pan and the comparison weights (→weight pieces, →standards) on the weight pan are interchanged after the first weighing (transposition weighing). This method compensates for the effect of the →lever error. It is used to check →equal-arm beam balances and weight pieces (e.g. →standard weights of relatively high accuracy). The method can only be used on equal-arm balances.

general approval

General approval is an approval of measuring instrument types for national certification (general national approval) or for →EC verification (general EC approval). In Germany, for example, the measuring instrument type must be stipulated in the annexes to the →Verification Ordinance [VO] as being generally approved, and must comply with the requirements and engineering standards specified for it in the VO. If the measuring instrument type does not comply with these requirements, a →type approval must be issued.

general clause

A general clause of an EC type approval allows a weighing instrument to be equipped with →modules that fulfill certain conditions. General clauses exist, for instance, for load cells and cash register systems.

German calibration service

→DKD

GLP

Abbreviation for →Good Laboratory Practice. →2004/9/EC, →2004/10/EC

GMP

Abbreviation for →Good Manufacturing Practice. →2003/94/EC

Good Automated Manufacturing Practice

Good Automated Manufacturing Practice (GAMP) is a technical sub-committee, known as a COP (Community Of Practice) of the International Society for Pharmaceutical Engineering (ISPE). The goal of the community is to assist companies in healthcare industries, including pharmaceutical, biotechnology, and medical devices, to achieve validated and compliant automated systems. GAMP publishes a series of Good Practice Guides for its members

on several topics involved in drug manufacturing. GAMP
was founded in 1991 in the United Kingdom to deal with
the evolving FDA expectations for →Good Manufacturing
Practice (GMP) compliance of manufacturing and related
systems. →Good Laboratory Practice

Good Laboratory Practice
Good Laboratory Practice (GLP) is a quality assurance
system that is concerned with the organizational process
and the conditions under which non-clinical health and
environmental safety studies are planned, performed, and
monitored. GLP is also concerned with recording, archiving,
and reporting of the tests.
The principles of Good Laboratory Practice are applied to the
non-clinical safety studies of test items that are included in
pharmaceutical products, pesticides, cosmetics, and veteri-
nary drugs, as well as food additives, animal feed additives,
and industrial chemicals. The purpose of testing these test
items is to obtain data about their properties and/or safety
for human health and/or the environment. The tests are initi-
ated by the state authorities responsible for the registration or
approval of products in the above-mentioned categories.
The OECD GLP Principles of 1997 were adopted as European law and formalized in European Directive →2004/10/
EC. Their verification is regulated in European Directive
→2004/9/EC. Both directives are implemented as national
law in the EEA and Switzerland.

Good Manufacturing Practice
Good Manufacturing Practice (GMP) is a quality assurance
system that ensures that products are consistently produced
and controlled to the quality standards appropriate to their
intended use and as required by the marketing authorization.
The prime focus is to avoid cross-contamination (in particular of unexpected contaminants) and mix-ups caused by, for
example, false labeling. GMP rules exist for various product
groups as medicinal products, medical devices, food, or
blood. For medicinal products, the rules are formalized in
European Directive →2003/94/EC. →Good Laboratory
Practice, →Good Automated Manufacturing Practice

gram
The gram (unit symbol "g") is the one thousandth part of the
→kilogram: 1 g = 0.001 kg.

gravimetric
→gravimetry

gravimetry

1. Method of quantitative analysis in which the mass, or a property that depends upon the mass, is determined by measuring the →mass; gravimetric determination. Gravimetry generally attains a greater accuracy of determination than, for example, →volumetry or →titration. It is therefore frequently used to verify other methods. →to weigh

2. Theory of the Earth's gravitational field and methods of determining →gravity. →acceleration due to gravity

gravitation

Gravitation is the term used to describe the mutual attraction that exists between masses irrespective of the material of which they are made. The force that arises depends on the masses of the bodies and their distance from each other (Fig. 76). The force of attraction F_G between two spherical bodies with masses m_1 and m_2 that are separated by a distance d (centers of gravity) is

$$F_G = G \frac{m_1 \cdot m_2}{d^2} \tag{2}$$

$G = 6.67 \times 10^{-11}$ N·m²/kg² (Newtonian gravitation constant) The forces that occur are relatively small, as shown by the following example: Two tankers each with a mass of 200000 t whose paths cross at a distance of 200 m are mutually attracted with a force of approximately

$$F_G = \left(6.67 \times 10^{-11} \frac{\text{Nm}^2}{\text{kg}^2} \right) \frac{\left(2 \times 10^8 \text{kg} \right)\left(2 \times 10^8 \text{kg} \right)}{\left(200 \text{m} \right)^2} \approx 67 \text{N}$$

which on Earth corresponds to a →weight force of approximately 6.7 kg. Gravitation is the principal source of terrestrial →gravity. →acceleration due to gravity

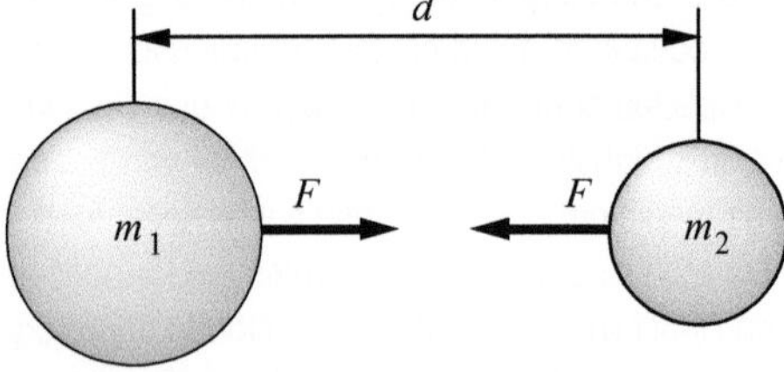

Fig. 76
Gravitation between two bodies

F_G: gravitational force
m_1, m_2: mass of bodies
d: distance between bodies
(see formula (2))

gravitational attraction
→gravitation

gravity

Gravity is the name given to the physical phenomenon that every body is caused by →gravitation to be attracted towards the Earth, that the body thus possesses weight, and that the body therefore exerts a →weight force on a support on which it rests. More specifically, the coefficient of proportionality g between the weight force F_G and the mass m of a body is referred to as gravity[7]. The gravitation formula [→gravitation, eq. (2)] can be rearranged so as to yield terrestrial gravity

$$g = \frac{F_G}{m} = G\,\frac{m_E}{R^2}$$

A mass for Earth of 5.97×10^{24} kg yields gravity at sea level ($R \approx 6371$ km) of

$$g \approx 9.81 \text{ N/kg}$$

In other words, every kilogram of mass is attracted towards the Earth with a force of slightly less than 10 newtons.
If the support on which the body rests is removed, the body is accelerated by the weight force towards the center of the Earth, i.e. vertically. It experiences the acceleration of free fall

$$a = g \approx 9.81 \text{ m/s}^2$$

which is identical to gravity.
Since different points on the Earth's surface are at different distances from its center of mass (due to the flattening of the poles and other physical features), the value of gravity is not constant. Gravity decreases with altitude by approximately 3×10^{-6} m/s^2 per meter. Gravity is reduced by the centrifugal acceleration of the Earth's rotating about its axis. This effect causes gravity to vary with respect to latitude. At sea level, gravity varies between the equator and the poles by approximately 0.5%.
Assuming the Earth to be a rotationally symmetrical body, gravity at any point on the Earth's surface can be calculated to a relative uncertainty $\Delta g/g$ of approximately 10^{-4} with the following formula:

$$g_0 = 9.780327\left[1 + 5.3024 \times 10^{-3}\sin^2(\varphi) - 5.8 \times 10^{-6}\sin^2(2\varphi)\right] - 3.086 \times 10^{-6}\,h$$

where

g_0 gravity according to GRS80 [m/s^2][8]
φ geographical latitude
h elevation above sea level [m].

The formula does not take account of gravitational anomalies (→Bouguer anomalies) due to local variations in density, which can cause a difference of up to 0.05% in the value of gravity.
Gravity is also affected to a limited extent by tidal forces[9]. The so caused relative fluctuations in gravity amount to some tenth of a part per million.

---

[7] also referred to as "gravitational acceleration"
[8] Geodetic Reference System 1980 (GRS80)
[9] these are the same forces that produce the ocean tides

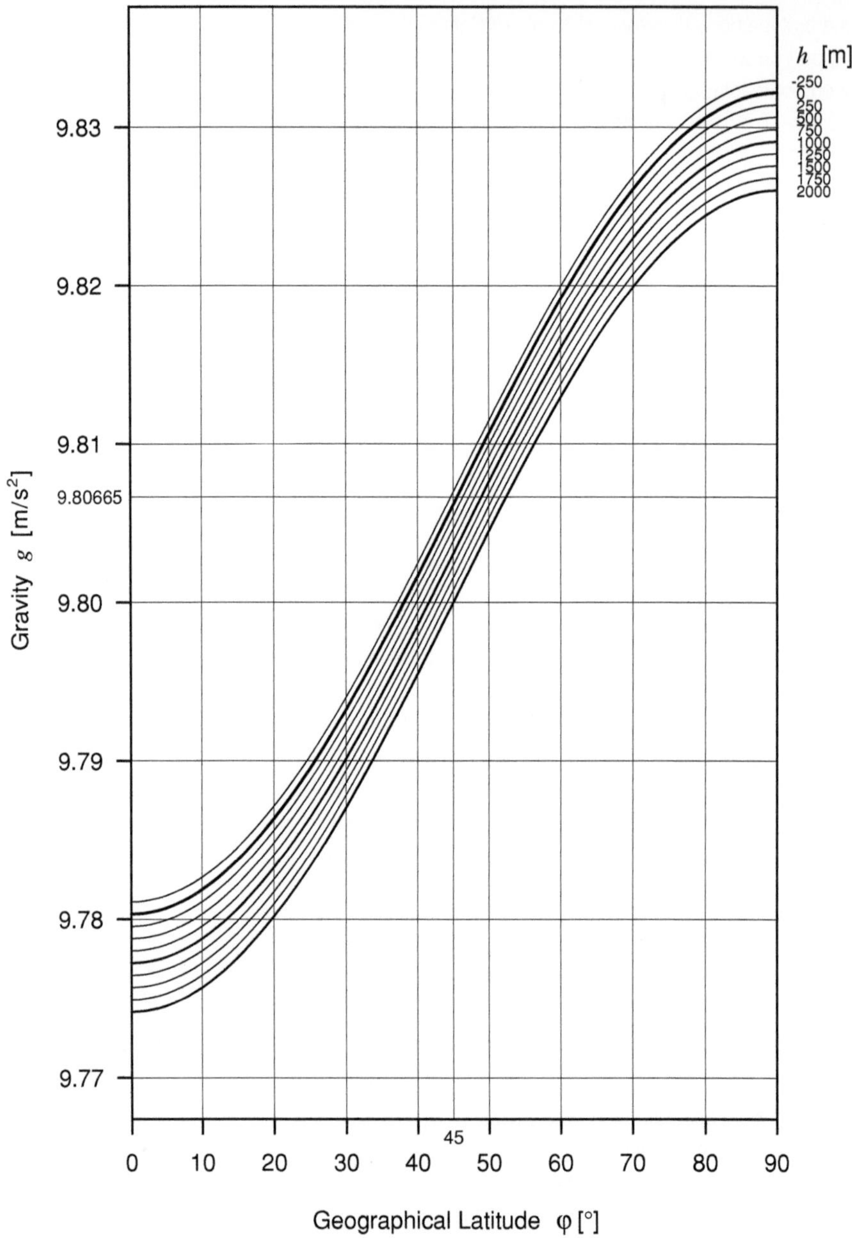

Fig. 77
Local gravity as a function of latitude and elevation above sea level (after GRS80[8])

φ: geographical latitude
g: gravity
h: elevation above sea level

[8] Geodetic Reference System 1980 (GRS80)

gravity-dependent weighing instrument
Weighing instrument whose weighing principle (→physical
weighing principle) does not measure the →mass of the
object itself but its →weight force (→force comparison).
If the →sensitivity of such a measuring instrument is not
pre-adjusted with a →reference mass (external or built-in) at
the →place of use, the weighing result is gravity dependent,
i.e. varies according to the place of use. →gravity, →stan-
dard gravity

Green M
Identification of a →non-automatic weighing instrument or
weighing instrument as defined in the →Measuring Instruments Directive by means of a black capital letter M on a
green background (Fig. 78). The Green M is affixed by the
manufacturer to complete instruments (not to →auxiliary
devices or →modules) to indicate conformity with the European →Directive on Non-Automatic Weighing Instruments or
→Directive on Measuring Instruments. In practice, the Green
M is also referred to as the →EC verification mark or metrology mark.

Fig. 78
Green M

gross value
Indication of the weight value (→weighing value) of a load
placed on a weighing instrument, with no →tare device
or →preset tare device in operation, often designated with
symbol G or B ([OIML R 76-1] T.5.2.1). →gross weight

gross weight
Total weight (→load) on the →platform of a weighing instrument, i.e. the weight of the weighed object (→net weight
or →sample) plus the weight of its container or packaging
(→tare weight).

guide
A mechanical connecting link (→parallel guide) to guide
→top-loading or →hanging load receptors so that they do
not tip over when loaded. The guide absorbs the guiding
forces that arise in such situations.

guided pan
→top-loading

GxP
Non official common collective abbreviation for
"GLP": (→Good Laboratory Practice);

"GMP": (→Good Manufacturing Practice);
"GCP": (Good Clinical Practice) and
"GDP": (Good Distribution Practice).

gyro load cell

Load cell in which the weight force of the weighed object is transferred to the axis of rotation of a rotating gyroscope. Instead of the axis of the gyroscope tilting, it undergoes displacement orthogonal to the force (so-called gyroscope principle, →physical weighing principle). This causes additional rotation (so-called precession) of the gyroscope. The frequency of precession of the gyroscope is proportional to the weight force acting (perpendicularly) on the axis of the gyroscope. →load cell

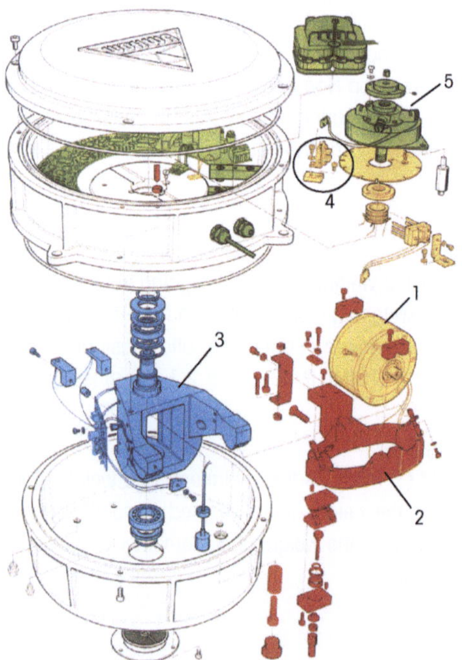

Fig. 79
Exploded-view drawing of a gyro load cell
The gyroscope (1) driven by the main motor is held in bearings in the inner gyroscope cage (2), which is itself suspended in the outer gyroscope cage (3). The weight force is transferred to the latter. The gyroscope and the gyroscope cage rotate under the effect of the weight force about the vertical axis (precession). The precession frequency is registered by a sensor (4). In addition, an auxiliary motor overcomes the precession friction that is present (so-called supporting motor, 5).

(Image by courtesy of Wöhwa Waagenbau GmbH, Pfedelbach, DE)

gyro measurement cell

→gyro load cell

gyro scale

An →electromechanical weighing instrument in which a →gyro load cell is used as measurement →converter.
→physical weighing principle

halogen lamp
Infrared heater in which the radiation is produced with the
aid of a halogen lamp. →dryer

hand scale
→Scale with a low maximum capacity which is held in the
hand when used. It is constructed as
- an →equal-arm beam balance,
- a simple sliding weight scale (→sliding weight balance)
  or
- a simple →spring scale.

hanger
Intermediate component located between the →levers and
the →load receptor or frame with →bearings, e.g. load
hanger, pendulum hanger.

hanging load receptor
→low-level load receptor

hanging pan
→low-level pan

hardware
1. Generic term for mechanical components such as
   screws, bolts, nuts, etc.
2. A term that applies to all of the mechanical and electrical
   components of a computer system or electronic circuit
   (e.g. printed circuit boards, transistors, integrated circuits, etc.) as well as to entire instruments. →software,
   →firmware

hierarchy of mass standards and weights
The definition and representation of the →unit of mass
→kilogram is tied to the →International Kilogram Prototype.
As a result, a hierarchical structure of mass standards and
weight pieces is adopted to achieve the highest possible
accuracy in mass determinations (Fig. 80). The structure
begins with the international prototype kilogram at the
→BIPM and ends with the →working standards used in
government and industrial metrology for trade and industry.

high-resolution
Non-technical term for weighments with a (relatively) high
→resolution, usually more than 10^4 to 10^5 scale intervals depending on the application. →analytical balance,
→weighing instrument of special accuracy

Fig. 80
Hierarchy of mass standards
in Germany

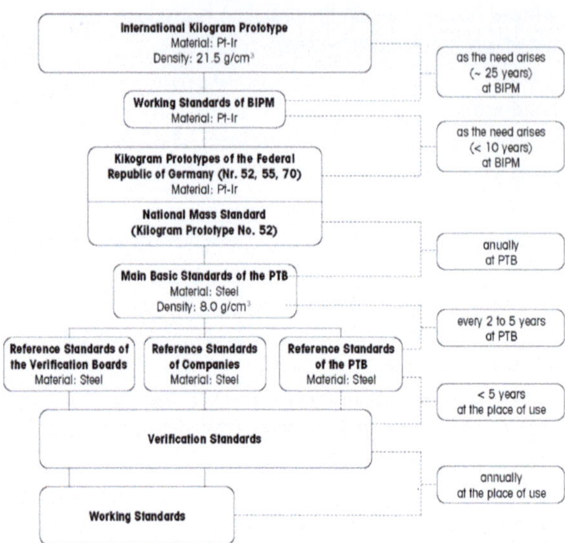

higher accuracy class
→Directive on Above-Medium Accuracy Weights

hopper scale
→Scale for hoppers, usually with →load cells in tension or compression between each of the hopper supports and an overhead structure or foundation, for →weighing or apportioning (→apportion) free-flowing bulk materials (Fig. 165). →bin scale, →tank scale

household scale
→Scale of low accuracy for use in private households (Fig. 81).

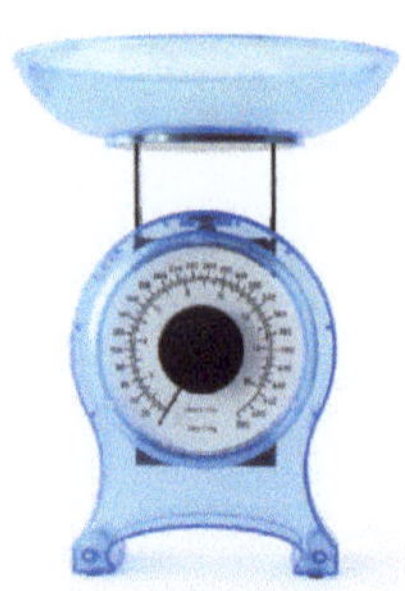
Fig. 81
Household scale

housing
A fixed cover that protects the sensitive parts of a weighing instrument and prevents inadmissible tampering with the instrument. →degrees of protection provided by enclosures

hump scale
→Scale for weighing railroad vehicles (individual uncoupled wagons) while passing over a prescribed weighing length in the track of the sorting hump. →automatic rail scale

hybrid weighing instrument
An →electromechanical weighing instrument in which the →weight force is reduced by a →lever system and transmit-

ted to one or more load cells. Frequently used for instruments whose mechanical →weighing-out devices have been replaced by →load cells, or for instruments with a high load limit and large →number of scale intervals. →bridge scale, →floor scale, →low-profile scale, →vehicle scale, →rail scale

hydrometer

Instrument for determining the →density of liquids (→density determination) or the concentration of dissolved substances (e.g. →Oechsle hydrometer) that takes the form of a glass tube with a scale that floats in the liquid whose density is to be determined (Fig. 82a). A hydrometer functions on the principle of →buoyancy, i.e. it floats higher or lower depending on the density of the liquid (Fig. 82b).

hydrostatic balance

Balance for determining the →density of a liquid by measuring the →buoyancy of a →sinker in the liquid, or for determining the density of a solid body in a liquid of known density (Fig. 83). →density determination, →Mohr-Westphal balance

hygroscopic weighing sample

Hygroscopic samples, for instance salts or filter papers, absorb moisture from the ambient air with the result that their mass constantly increases. Because of this, and depending on the →resolution of the weighing instrument, when such substances are weighed, no stable →measurement value can be expected. It is advisable to use weighing containers that have a narrow neck and are closed with a lid. →influence of moisture

hysteresis [11]

The phenomenon that a measuring instrument indicates two different →measurement values for the same →measurand, depending on whether the measurand is increasing or decreasing. This results in a split →characteristic curve: the lower curve is for the increasing, the upper for the decreasing measurand (Fig. 84). Hysteresis can be compensated for (→hysteresis compensation device). If hysteresis is overcompensated, the characteristic curve reverses its course. →hysteresis deviation

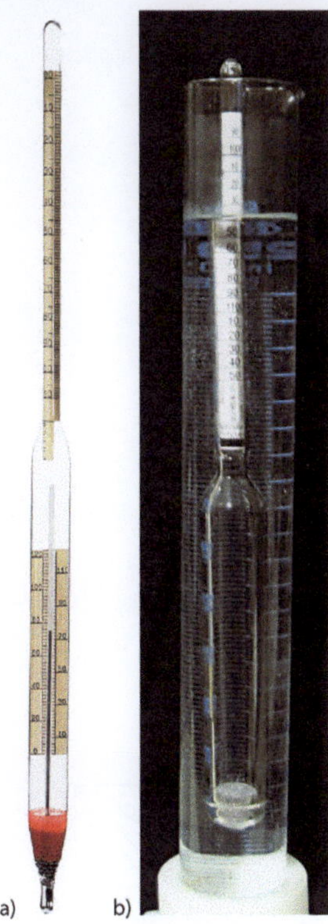

Fig. 82
a) Hydrometer with integrated thermometer; the density scale is located at the upper end of the narrow glass tube, the temperature scale at the lower end in the displacement body. (Image by courtesy of Cole-Parmer Canada Inc., Montreal, Canada)

b) Submersed hydrometer
(Image from Wikimedia Commons (Author: Han-Kwang Nienhuys) is available under the GNU license [10].)

---

[10] GNU Free Documentation License: http://www.gnu.org/licenses/fdl.txt
[11] hysteros (Greek): lagging

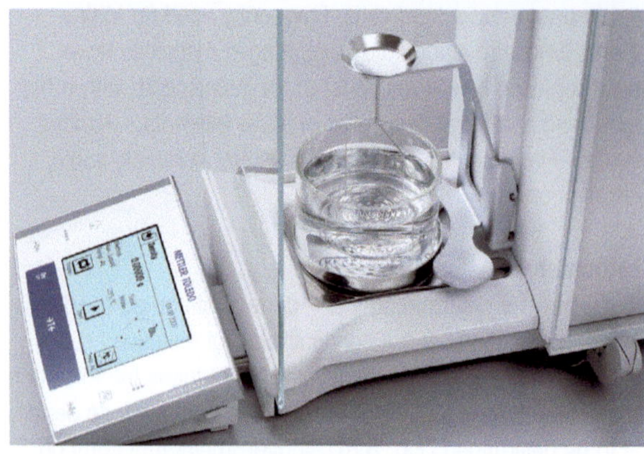

Fig. 83
Density determination apparatus

hysteresis compensation device

A device or measure to compensate the →hysteresis of a →weighing instrument, e.g. special spring arrangements on a →spring scale, or computerized compensation of the hysteresis of an →electromechanical weighing instrument.

hysteresis deviation

Difference between two →weighing values for the same load, one obtained with increasing load and the other obtained with decreasing load (Fig. 84).

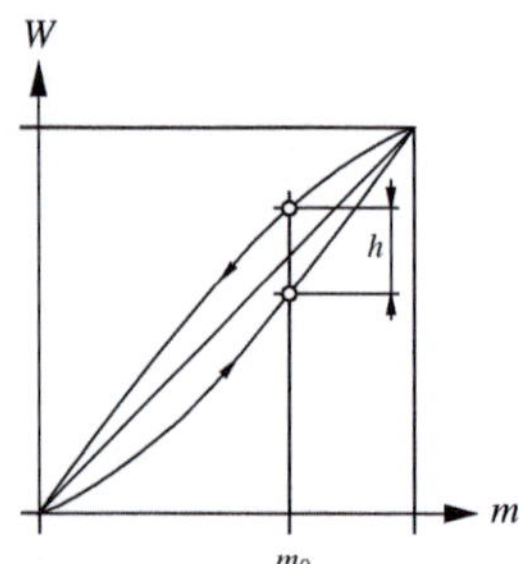

Fig. 84
Characteristic curve of load m and weighing value W of a measuring instrument with hysteresis h at load m_0

identification mark

A marking, usually a manufacturer's number, that is applied to the →main devices of the weighing instrument consisting of separate units to ensure their unambiguous association ([OIML R 76-1] 7.1.2).

IEC

Abbreviation for →International Electrotechnical Commission.

IEC 60529

IEC standard "Degrees of protection provided by enclosures (IP Code)", →degrees of protection provided by enclosures

ILAC

→International Laboratory Accreditation Cooperation

inclination

Angle between the →axis of action of the →weighing instrument and the vertical. →tilt

inclination error

A deviation in the sensitivity resulting from an →inclination of the weighing instrument. The sensitivity S decreases in proportion to the cosine of the angle of →tilt α. Therefore, for small angles, the following approximation may be used

$$\frac{\Delta S}{S} \approx -\frac{1}{2}\alpha^2$$

S sensitivity [1]
ΔS sensitivity offset [1]
α angle of tilt [rad]

inclination range

Scale range of a →deflection weighing device, in most cases the automatic →self-indication capacity of a →deflection balance indicated by the readout device.

inclination sensor

Device that measures the deviation of the →axis of action of the measuring instrument from the vertical (→inclination) (Fig. 85). →automatic inclination sensor, →level indicator

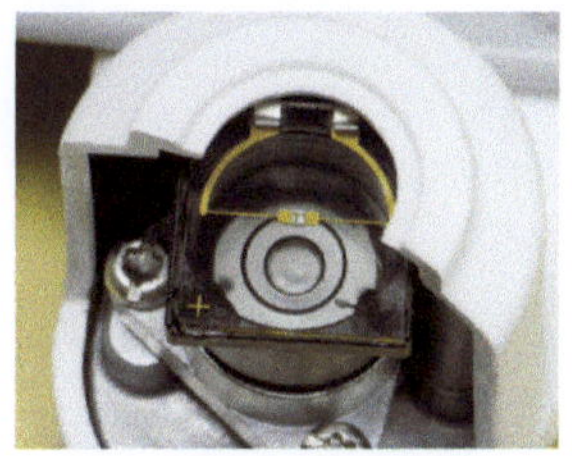

Fig. 85
Cutaway view of an optical inclination sensor

inclination test

A test mode with weighing instruments that do not have a fixed location and are also not freely suspended. The test indicates whether an inclination or tilting of the instrument results in a change in the weighing value. Indicating devices for the inclination are, for example, the →level indicator and the →plumb line. →inclination

inclinometer
→inclination sensor

indication
Value that is represented on a display and can be read (→measurement value) ([OIML R 76-1] T.1.3).

industrial scale
→Scale for industrial use, e.g., →weighbridge, →truck scale or →platform scale, verifiable (where applicable) to →accuracy class ⑪ or ⑫ (or III L in the USA). →bridge scale

influence of electrostatics
→electrostatic influence

influence of environment
→environmental influence

influence of humidity
→influence of moisture

influence of moisture
→Air humidity results in an adsorbed film of moisture on virtually all surfaces. An equilibrium between the adsorption film and the surrounding humidity arises. Moisture differences between the weighing sample and the air in the weighing chamber can thus lead to mass changes in the weighing sample and hence cause drift in the weighing instrument display. Corrective measures include the use of clean, dry weighing vessels. In addition, instead of cork or cardboard surfaces, which may absorb or release a considerable amount of moisture, non-hygroscopic auxiliary equipment (e.g. →triangular support) should be used. →Weight pieces are also affected, and should therefore always be acclimatized before use, especially for accurate calibrations.
The air humidity also affects the →air density.

influence of temperature
→temperature influence

influence quantities
1. Quantity that is not the measurand but that affects the result of the measurement. ([GUM] B.2.10)

2. Quantity that, in a direct measurement, does not affect
   the quantity that is actually measured (→measurand),
   but affects the relation between the →indication and the
   →measurement result. ([VIM:2008] 2.52)

Variables and conditions that may affect the normal working
of a weighing instrument include, for example, the →ambi-
ent temperature, →air humidity, and →air pressure; also
→electrostatic charges, magnetic fields, electric power supply networks, →vibrations, mechanical stresses, →tilt, and
→air buoyancy, as well as feeding, filling and emptying
devices that are connected to the instrument.

infrared dryer

→dryer

ingress protection

→degrees of protection provided by enclosures

initial verification

→Verification of a measuring instrument which has not been
verified previously ([VIML] 2.15), contrary to →subsequent
verification. →EC verification

initial zero-setting device

A device that is used to set the indication automatically to
zero when the weighing instrument is switched on and before it is ready for use ([OIML R 76-1] T.2.7.2.4).

initial zero-setting range

→Load range within which the →display device is capable
of being set to zero after switching on the →weighing instrument. ([OIML R 76-1] A.4.2.1.1)

inscriptions

Specifications and designations used for a more detailed
description of a weighing instrument on its →data plate and
in its operating instructions. The inscriptions may include
the name of the manufacturer and the model, serial number,
maximum capacity, operating voltage and power supply frequency. There may also be references to type data, approval
data, instructions for air buoyancy correction, prohibition of
use at points of sale, or possible uses and applications of
the weighing instrument, etc. (Fig. 86).

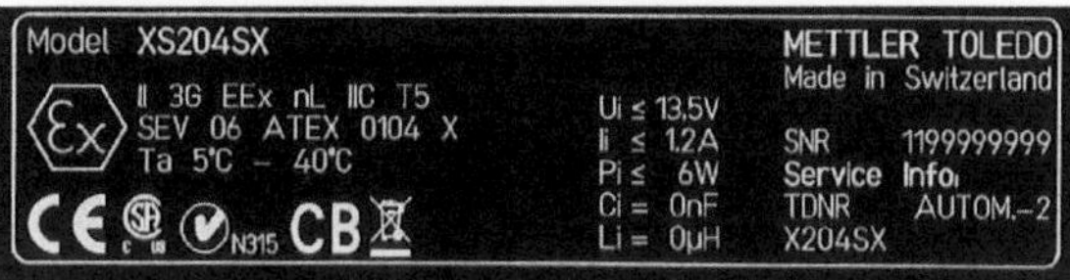

Fig. 86
Example of inscriptions:
Model name (top left),
manufacturer's name (top right),
serial number (SNR),
type-specific information (TDNR),
specific information about explosion
protection (Ex, middle left),
including electrical parameters
(center) and safety parameters
including CE mark (bottom left).

inspection

Test performed by the →Weights and Measures authorities to determine whether a measuring instrument attested as verified still operates within the →maximum permissible error in service and thus still satisfies the requirements for its certification (type approval).

installation of weighing instruments

→High-resolution weighing instruments, such as →precision balances (→weighing instrument of high accuracy) and →analytical balances (→weighing instruments of special accuracy), should be installed on (→place of installation) a →weighing table in a vibration-free room with the least possible fluctuations in temperature and humidity (possibly with air conditioning) (→environmental influence). There should be only one entrance to this room so that it cannot be used as a passageway and there are no air currents. The corners of the room are particularly suitable as workplaces since these are the most rigid points of a building. It is important to avoid exposure of the instrument to radiant heat from direct sunlight, radiators, etc., to air currents from windows, cold walls or air conditioning, or to vibrations from movement or rotation of the →support (Fig. 87a and 87b).

Fig. 87
Installation of high-resolution weighing instruments
a) To the extent possible, the instrument should not be installed near to corridors or windows, and preferably in the corner of a room.
b) The instrument should be placed on a stable support. Suitable surfaces are
c) a benchtop fastened to a stable wall, or
d) a stone bench.

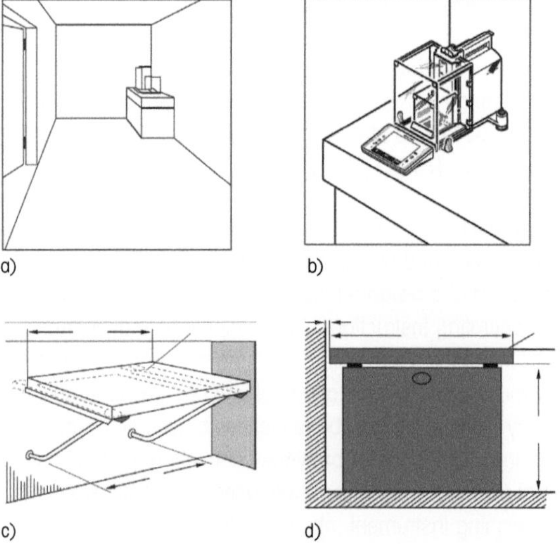

Solid stone slabs are particularly suitable for the construction of weighing tables. The slabs are either fastened to the wall by means of brackets (Fig. 87c) or placed on two massive supports that stand on the ground, preferably with damp-

ing elements between the slab and the supports (Fig. 87d).
Fastening to both the wall and floor is not recommended
since the vibrations of both are then transmitted to the instrument. Heavy objects should not be placed or left on the
table.
A windowless room with artificial lighting is most suitable.
The lights must be installed sufficiently far from the weighing
table. Filament lamps should be avoided; fluorescent light
tubes or other light sources which emit little heat are prefer-
able.

Installation Qualification
Part of →Equipment Qualification (EQ). The Installation
Qualification (IQ) verifies that the instrument is delivered as
specified, that it is correctly installed, and that the environment is suitable for its operation.

integration time
The time required by an (electronic) measuring instrument to
form a measurement value. The expression derives from the
time during which the partial measurement values output by
a measurement →converter are added (integrated) to form
a measurement value with sufficient resolution or stability. In
the case of →digital filters, the integration time is not normally fixed, but comprises an increasing and decreasing weight-
ing with which completed partial results are summed to form
the momentary measurement value. →measurement time

integration time extension
Device that adapts the →integration time to the type of dis-
turbances that occur.

interchange weighing method
→Gaussian weighing method

interface
Point of contact or connection between two data transmission devices. The term "interface" embraces all of the characteristics that describe its physical, electrical, and logical
functions at the point of transfer. The characteristics include
those of the plug, the pin assignments, the voltage and current levels, the data format and coding, as well as the data
and commands that are transferred. →protected interface

interference quantities
→influence quantities

International Electrotechnical Commission

International organization with permanent headquarters in Geneva. The International Electrotechnical Commission (IEC) develops and publishes standards for electrical, electronic, and related technologies (www.iec.ch).

International Kilogram Prototype

Synonym for the →International Prototype of the Kilogram.

International Laboratory Accreditation Cooperation

International cooperation of laboratory and inspection accreditation bodies to help remove technical barriers to trade. The aim of the ILAC arrangement is the increased use and acceptance by industry as well as regulators of the results from accredited laboratories and inspection bodies, including results from laboratories in other countries. (www.ilac.org)

International Organization for Legal Metrology

International organization with permanent headquarters in Paris (→BIML). The mission of the International Organization for Legal Metrology (OIML) is the international harmonization of the administrative and technical regulations for measurement methods and →measuring instruments in the field of legal metrology. For this purpose, the organization issues recommendations and documents (→OIML Recommendations and Documents) for individual measuring instruments (www.oiml.org).

International Organization for Standardization

International organization that has its headquarters in Geneva. The International Organization for Standardization (ISO) undertakes the international standardization of terminology, measurement methods, →tolerances, etc. in the industrial field (www.iso.org).

International Prototype of the Kilogram

Definition and representation of the mass unit (→mass). This prototype is kept at the International Bureau of Weights and Measures (→BIPM), Pavillon de Breteuil, Sèvres, France (→hierarchy of mass standards and weights) (Fig. 88). The International Prototype of the Kilogram (→prototype) was created and sanctioned by the →CIPM in 1889. It is a cylinder with both height and diameter of approximately 39 mm. It is made of an alloy of 90% platinum and 10% iridium with a density of 21500 kg/m^3. The

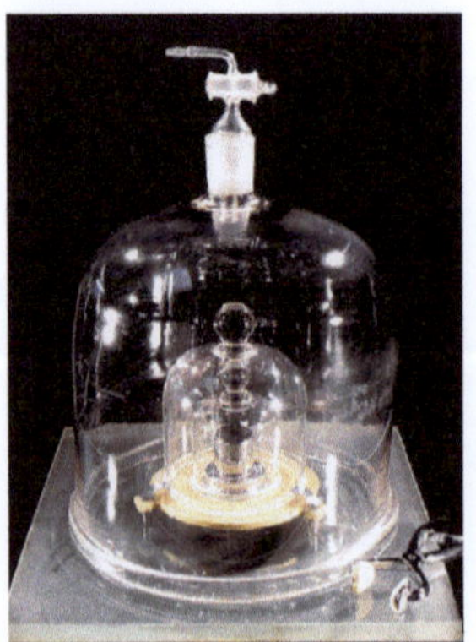

Fig. 88
Pt-Ir facsimile of the International Kilogram Prototype at the BIPM in Sèvres, Paris
(Image by courtesy of BIPM, Sèvres, France)

kilogram is the last of the base units of the →International
System of Units to be defined by an artefact, i.e., a manufactured object.

International System of Units

International system of units, also called SI (abbreviation
of the French name 'Système international d'unités'), that
embodies the →metric system of units and is the most
widespread system of physical and chemical units. The
International System of Units has seven base units:

Dimension	Name	Unit Symbol
length	meter	m
mass	kilogram	kg
time	second	s
electric current	ampere	A
thermodynamic temperature	kelvin	K
quantity of substance	mole	mol
luminous intensity	candela	cd

Other units are derived by means of simple relationships,
e.g.:

Dimension	Name	Unit Symbol
area	square meter	m^2
volume	cubic meter	m^3
speed	meter per second	m/s
acceleration	meter per second per second	m/s^2
force	kilogram meter per second per second	$kg \cdot m/s^2$
wave number	reciprocal meter	1/m
density	kilogram per cubic meter	kg/m^3
electric current density	amperes per square meter	A/m^2
magnetic field strength	amperes per meter	A/m
concentration of substance	mol per cubic meter	mol/m^3
specific volume	cubic meter per kilogram	m^3/kg
luminous density	candela per square meter	cd/m^2

Various derived units have been given specific names and specific unit symbols, e.g.:

Dimension	Name	Unit Symbol	Expressed in other SI units	Expressed in SI basic units
frequency	hertz	Hz		$1/s$
force	newton	N		$kg \cdot m/s^2$
pressure, mech. tension	pascal	Pa	N/m^2	$kg/(m \cdot s^2)$
energy, work	joule	J	$N \cdot m$	$kg \cdot m^2/s^2$
power, energy flow	watt	W	J/s	$kg \cdot m^2/s^3$
Electric charge	coulomb	C		$A \cdot s$
el. potential, el. voltage, electromotive force	volt	V	W/A	$kg \cdot m^2/(A \cdot s^3)$
el. capacitance	farad	F	C/V	$A^2 \cdot s^4/(kg \cdot m^2)$
el. resistance	ohm	Ω	V/A	$kg \cdot m^2/(A^2 \cdot s^3)$

SI units are also referred to as →metric units, as opposed to →nonmetric units.

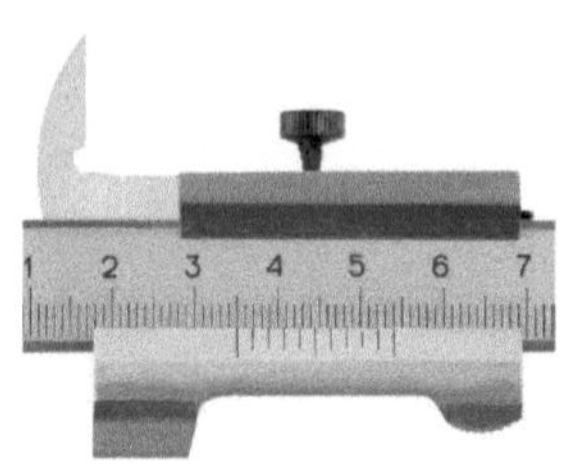

Fig. 89
Vernier

interpolation device
Device that is rigidly connected to the equilibrium indicator that makes it possible to subdivide an evenly divided scale into finer parts without manual intervention (vernier, Fig. 89) ([OIML R 76-1] T.2.5.2).

intervention limit
→control limit

invariability
Obsolete term for →repeatability

IP code
→Degrees of protection provided by enclosures in accordance with IEC 60529.

IP protection
IP is the abbreviation for Ingress Protection. →degrees of protection provided by enclosures

ISO
Abbreviation for →'International Organization for Standardization'.

ISO 17025
International standard "General Requirements for the Competence of Testing and Calibration Laboratories" which describes the requirements for calibration and testing laboratories. The requirements cover the areas of personnel, technical infrastructure and organizational structure. This insures that the product or service of the laboratory is competently produced or provided and satisfies the requirements.

joint
Movable connection for mutual guidance of two mechanical
components or for the transmission of guiding forces between them. →flexible joint, →pivot joint, →flexible coupling, →cross-flexed spring joint

joint flexure
→flexible joint

k

→expansion factor

kg

Unit symbol for the mass unit →kilogram.

kilogram

The kilogram (unit symbol "kg") is the unit of mass in the
→International System of Units. It is one of the seven base
units of this system. →unit of mass

kilogram prototype

→International Kilogram Prototype

knife-edge

That part of the →knife-edge bearing that makes contact
with the →pan, a.k.a. flat or bearing. →design and function
of a mechanical balance

knife-edge angle

The angle enclosed by the wedge-shaped part of the planes
that form the →knife-edge; usually greater than 90° if made
of hard or hardened and consequently brittle materials.

knife-edge bearing

Suspension of the moving parts (e.g. levers) of a balance by
means of →knife-edge, a.k.a. pivot, and →pan, a.k.a. flat
or bearing (Fig. 90). →design and function of a mechanical
balance

knife-edge plane

Plane through the parallel knife-edge lines of a balance
lever.

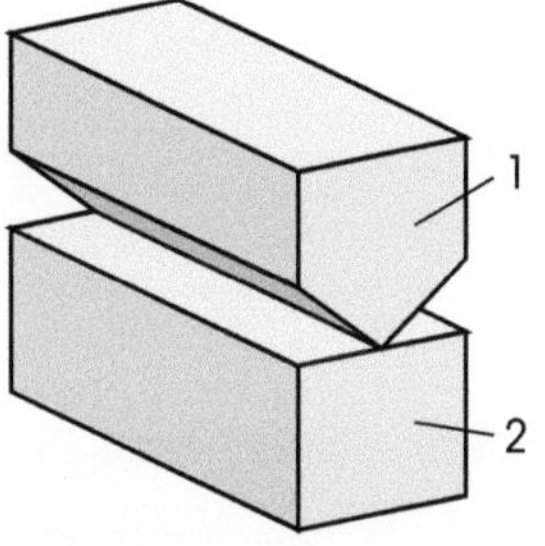
Fig. 90
Knife-edge bearing

1: knife-edge or pivot
2: pan, flat or bearing

label printer

Printer that prints →weighing values and other data on
labels with a predefined format. Usually a component of a
→price marker scale.

labels

→inscriptions

laboratory balance

Designation for balances that are mainly used in laboratories and usually referring to →precision balances or →ana-
lytical balances.

Fig. 91
A selection of laboratory balances:
Analytical balance (0.0001 g), precision balances with readabilities of
0.001 g, 0.01 g and 0.1 g (from left
to right).

legal metrology

Part of metrology relating to activities which result from
statutory requirements and concern measurement, units of
measurement, measuring instruments and methods of measurement and which are performed by competent bodies.
The scope of legal metrology may be different from country
to country. The competent bodies responsible for all or part
of these legal metrology activities are usually called legal
metrology services. ([VIML] 1.2)

legal metrology requirements

Regulations that must be fulfilled by a measuring instrument to be used in applications subject to legal metrology
(→compulsory verification). For →non-automatic weighing instruments (NAWI), these are European Directive
→2009/23/EC and →European Standard EN 45501, for
→automatic weighing instruments European Directive
→2004/22/EC, and for →weight pieces, European Directive
→71/317/EEC and →74/148/EEC.
In Germany, the respective regulations are stipulated in
the →Verification Ordinance [VO], General Section, and in

Appendix 9 for non-automatic weighing instruments, Appendix 10 for automatic weighing instruments, and Appendix 8 for weight pieces.

legally relevant parameter
Parameters and data of verified measuring instruments or →modules. ([OIML R 76-1] 2.8.2)

legally relevant software
Programs, data, and type-specific parameters of the →measuring instrument or →module that contain or fulfill verified functions ([OIML R 76-1] T.2.8.1).
Examples: Final →measurement values (gross, net, tare, tare input, decimal sign, unit), display of the weighing range and of the →load receptor (if multiple load receptors are present), →software identification.

letter scale
→Scale for weighing letters, small parcels, and printed matter.

level
→inclination

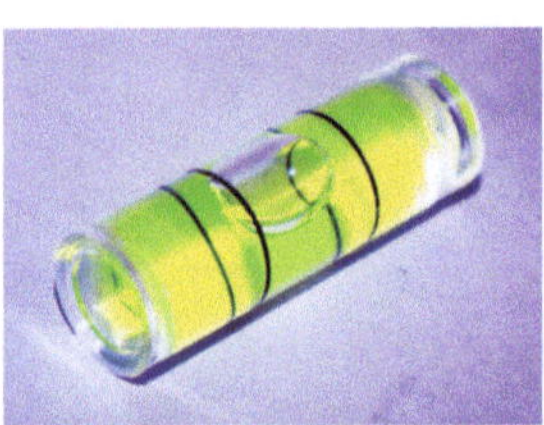

Fig. 92
Tubular level indicator
(Image by courtesy of R. Bormann & Sohn, Rabenau-Lübau, DE)

level indicator
Device that indicates the →inclination. A level indicator usually comprises a sealed container (of glass or clear plastic) that is filled with liquid on which a gas bubble floats (Fig. 92). Depending on the application, different shapes such as tubular, cross-shaped or →circular level indicators are used. →inclination error, →level

level sensor
→inclination sensor, →level indicator

level, to
Adjusting a weighing instrument to its →reference position (usually horizontal) so that its →axis of action is parallel to the vertical. This usually means setting the housing of the weighing instrument horizontal. Certain types of weighing instruments, particularly those considered to be portable, must be fitted with either a →level indicator or a →plumb line. →inclination error, →level, →leveling device

LEVEL-MATIC®
→Load receptor that automatically centers a load placed eccentrically on the weighing pan to avoid →eccentric load

deviations (Fig. 93). This is achieved by the load receptor
(→pan or →platform) having the form of a spherical segment and being movably supported on spherical elements.

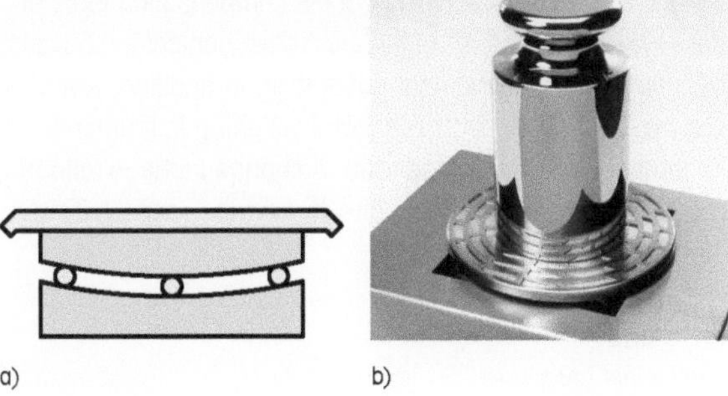

a)   b)

Fig. 93
Device for avoiding eccentric load deviations by means of a LEVEL-MATIC®
load receptor
a) operating principle;
b) a LEVEL-MATIC® load receptor

leveling device
Device to align (→level) a weighing instrument in its →ref-
erence position (usually horizontal), for example with leveling screws ([OIML R 76-1] T.2.7.1). →level indicator

leveling screws
Screws, usually on the baseplate or →frame, that are used
to align the weighing instrument in its →reference position.

levelness compensation
A device that automatically compensates for a change in
→measurement value caused by a change in the levelness
of the weighing instrument.

lever
A rigid body used to transfer forces or torques. When used
as balance levers, they can usually be rotated about a
horizontal axis to compare forces acting in a vertical direction during weighing. The lengths of the levers are defined
by →pivot joints. Levers can be arranged in parallel or in
series.
A distinction must be made between two-arm levers
(Fig. 94a) in which the pivot point (fulcrum) is located
between the points of application of the load and the compensating force, and the one-arm lever (Fig. 94b), in which
the pivot joint is located outside these points of application.
From a constructional point of view, levers can be designated as single levers, triangular levers, revolving-type levers
or angular levers. From the functional aspect, weighing-out

® Registered trade mark of METTLER TOLEDO

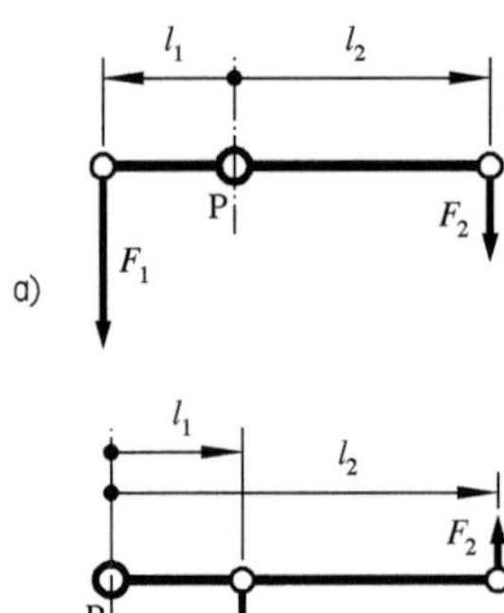

Fig. 94
Levers:
a) double-arm lever;
b) single-arm lever

P: fulcrum
l_1, l_2: lever arms
F_1, F_2: lever forces

levers, poise, inclination, dial weight, load, connecting, and test levers can be found.

Levers are used to adapt, usually meaning reduce, the weight force of the item being weighed, or of the reference weight, to the restricted range of the compensation force of the →weight pieces, or of the electromechanical →converter, by means of →mechanical advantage. In addition, levers can be used on electromechanical weighing instruments to compensate the →dead load. A change in the →ratio of mechanical advantage acts directly on the →sensitivity of the balance.

lever arm
→effective lever arm

lever arm, effective
→effective lever arm

lever chain
→lever system

lever error
A deviation of the →ratio of mechanical advantage of a →lever or →lever system from its →nominal value. The Gaussian and Borda weighing methods (→substitution weighing) eliminate the lever error.

lever group
→lever system

lever ratio
→ratio of mechanical advantage

lever system
1. Name given to the lever of a balance.
2. Name given to several levers arranged in sequence (chain of levers) or side by side (group of levers).

leverage
→ratio of mechanical advantage

Lim
Abbreviation for "load limit" (→maximum safe load).

limit switch
A mostly mechanical or →electronic device on weighing instruments that triggers certain control functions when a

(preselected) limit value of the load or measurement value has been reached.

limit value of inclination
Maximum permissible →inclination of a weighing instrument that is reached when the air bubble of the →level indicator has moved so far away from its centered position that it touches a corresponding mark (e.g. ring, line) ([OIML R 76-1] 3.9.1.1). With common level indicators, this occurs at an inclination of 0.1%.

limit value of tilt
→limit value of inclination

limits of measurement errors
→error limits

linearity
Ability of a weighing instrument to follow the linear relationship (Fig. 95) between a load m and the indicated →weighing value W. →sensitivity; compare: →nonlinearity

linearity deviation
→nonlinearity

linearization
Device or measure that eliminates the →nonlinearity of a →measuring instrument and thereby produces a linear →characteristic curve. The characteristic curve of a weighing instrument is corrected, i.e. straightened by mechanical or electronic means or with the aid of an algorithm that is executed by the →signal processing unit. The linearity deviation is determined by weighing external or built-in →reference weights. →FACT

liquid thermometer
Thermometer that determines the temperature from the expansion of a column of liquid (e.g. alcohol, mercury).

LNE
Abbreviation for 'Laboratoire national de métrologie et d'essais'. French →national metrology institute with headquarters in Paris (www.lne.fr).

load
General term used to refer to an object that is exerting a →weight force. An object that is placed on the →load recep-

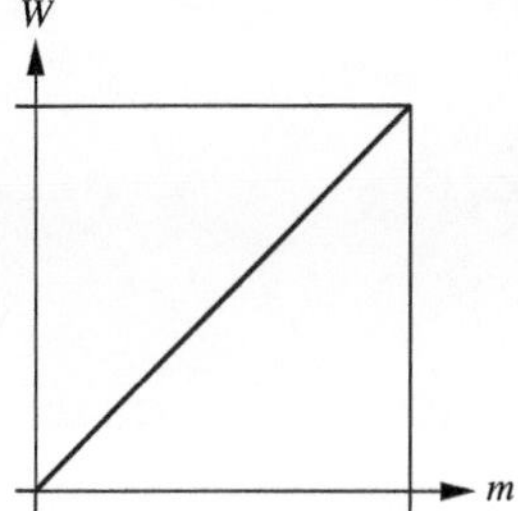

Fig. 95
Linear characteristic curve of a weighing instrument between load m and weighing value W

a)

b)

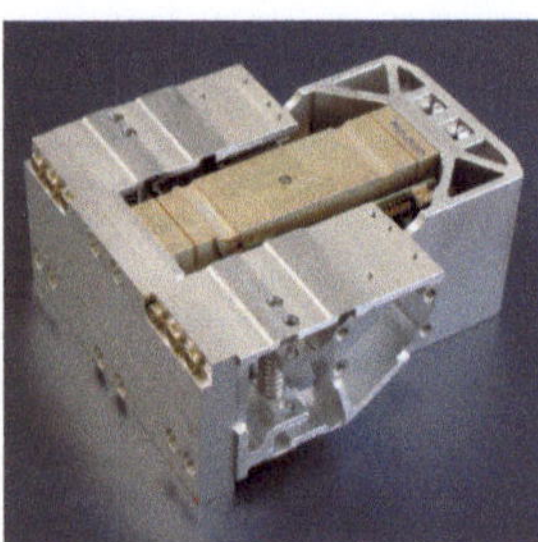

c)

d)

Fig. 96
Load cells
a) strain gage load cell;
b) string load cell;
c) EMFC load cell;
d) gyro load cell

(Images by courtesy of Digisens AG, Murten, CH (Fig. 96b) and Wöhwa Waagenbau GmbH, Pfedelbach, DE (Fig. 96d))

tor of a weighing instrument is generally referred to as the load because its →weight effect is essential for the weighing operation. →loading

load cell
Electromechanical →measurement transducer for determining mass, in which the →weight force exerted by the →weighed object is converted into an electrical →signal, e.g. →strain gage load cell (Fig. 96a), →string load cell (Fig. 96b), →EMFC load cell (Fig. 96c), →gyro load cell (Fig. 96d) ([OIML R 76-1] T.2.2.1). →physical weighing principle

load compensation
Establishment of equilibrium between the →weight force of the →load and the compensating counterforce of the →weighing-out device.

load drift
→Drift of the →measurement value that is caused by, and occurs after loading or unloading the weighing instrument.

load lever
→Lever of a composite balance that supports the →load receptor and transfers the force to the →measurement transducer.

load limit
Non-technical term for →weighing range, →maximum capacity, or →maximum rated load.

load pan
Name given to the pan-shaped →load receptor on weighing instruments with a low →maximum capacity.

load range
Range within which a weighing instrument can be loaded and function correctly, usually extending from the →zero point (unloaded weighing instrument) to the →maximum capacity. →Mass comparators may require a →minimum load. →weighing range, →nominal load range

load receptor
That part of the weighing instrument that carries or accommodates the →load, e.g. →weighing pan, →load pan, load hook, →platform, →bridge or container (→weighing container) ([OIML R 76-1] T.2.1.1).

load relief device

A device on a balance by means of which the frictional connection between the →load receptor and the →weighing-out
device can be disconnected. →locking device

load, eccentric

→eccentric load, →load

loading

1. The entirety of objects on the platform that exert a force
   onto the weighing instrument.
2. →Nominal load, →maximum capacity, →minimum
   load, →weighing range, →self-indication capacity,
   →maximum tare effect, and →maximum rated load are
   characteristic loads of a weighing instrument.
3. Mechanical stress of a weighing instrument resulting
   from the load placed upon it.
4. If a test load, or more specifically its center of gravity,
   is placed on the →load receptor asymmetrically, this is
   referred to as eccentric loading or →eccentric load.

local gravity

The value of →gravity at the →place of installation. The local value of gravity is caused mainly by →gravitation, and
therefore depends on the elevation (distance from the center
of the Earth) and to a much lesser extent on local anomalies
(→Bouguer anomaly). Gravity is reduced by the centrifugal
acceleration that arises from the rotation of the Earth, which
depends on geographical latitude. Because of these factors,
terrestrial values of gravity can vary by up to 0.5%. There
are additional twice-daily variations to local gravity caused
by tidal forces; however, these are so small ($< 10^{-6}$) that
they do not affect common weighing processes.
Because of these spatial variations, the →weight force of a
body becomes correspondingly greater or less. As a conse-
quence, the →sensitivity of →gravity-dependent weighing
instruments must either be preset for the →place of use, or
adjusted on site (→sensitivity adjustment).

locking

→locking device, →automatic release

locking device

A device used especially in high-resolution mechanical
weighing instruments to separate the →knife-edges from the
→pans and/or lock the lever, connecting hanger, and pan
so that they are protected, for example, during loading or

transportation ([OIML R 76-1] T.2.7.6). →design and function of a mechanical balance, →automatic release

long-term stability
→stability

long-term storage of measurement data
Storage of →weighing results and associated data from a weighing operation for subsequent applications that are subject to legal metrology requirements (e.g. →printout of the weighing results on the invoice to a customer at a later date). ([OIML R 76-1] T.2.8.5)

low resolution
Non-technical term for weighments with a (relatively) low →resolution, usually with a →number of scale intervals less than 10^4. (Compare: →high-resolution, →weighing instrument of medium accuracy, →weighing instrument of ordinary accuracy)

Low Voltage Directive
European Directive for electrical equipment with a nominal voltage between 50 and 1000 V for alternating current, and between 75 and 1500 V for direct current. The directive defines technical requirements for these devices so that when correctly installed, maintained, and used as intended, they do not present a hazard to the safety of humans or domestic animals, or to the preservation of property. Non-technical term that is frequently used in this connection is →electrical safety. The Low Voltage Directive is implemented as national law in the EEA and Switzerland. →2006/95/EC

low-level load receptor
→low-level pan

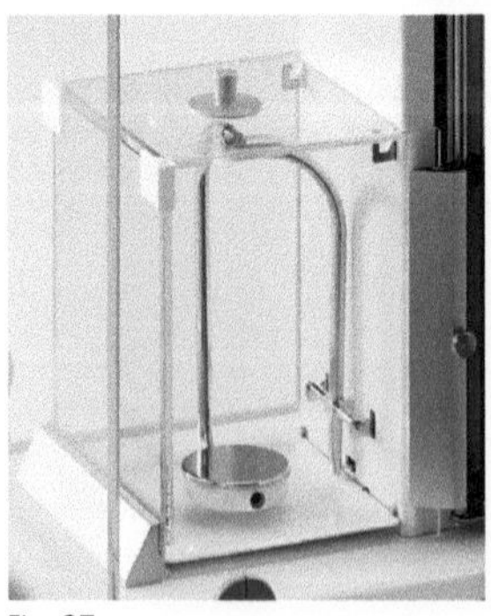
Fig. 97
Hanging (low-level) load receptor

low-level pan
Designates the type of construction of a balance in which the →load receptor is freely suspended on the pivot point (fulcrum) of the load arm or at the point where the force enters the →load cell (Fig. 97). Irrespective of where the weighing sample is placed on the load receptor, the common center of gravity of the load receptor and weighing sample is always vertically below the point of suspension, which prevents eccentric load errors (→eccentric load). Depending on the type of suspension, after the low-level load receptor has been charged with the sample to be weighed it may oscillate and thus prolong the measure-

ment time (→pan brake). →below-the-balance weighing
(compare: →top-loading)

low-profile scale
→Bridge scale that is not permanently installed and has a
flat →load receptor (Fig. 98).

Fig. 98
Low profile scale (capacity up to 3 t)

lumpiness of the weighing sample
Weighed material (filling material) is classified into different
fill groups according to the average piece weight compared
with the fill weight in question. In metrology, three fill groups
with limit values for the average piece weight are defined as
a function of the fill weight.

LVD
Abbreviation for →'Low Voltage Directive'.

machine
A totality of parts or devices that are joined to each other and
at least one of which is movable. In the European Union, the
requirements for machines are regulated by the →European
Machinery Directive. →Machinery Directive

Machinery Directive
This European Directive regulates the measures to ensure
that machines and safety components may only be placed
on the market and put into operation provided that, with appropriate installation and maintenance, and when used as
intended, they do not present a hazard to human health. To
this end, the manufacturer must ensure that the machines
comply with the so-called basic essential health and safety
requirements that are listed in the directive. The directive is
implemented as national law in the EEA and Switzerland. If
the hazards are mainly of an electrical nature, the Machinery
Directive is not applicable, and the instrument is governed
exclusively by the Low-voltage Directive →2006/95/EC.
→2006/42/EC (→98/37/EC)

macroanalytical balance
→Analytical balance designed for macroanalysis that has
a maximum capacity of approximately 100…200 g and a
readability of 0.1 mg.

magnetic damping
→damping systems

magnetic suspension balance
Balance with a →peripheral device that allows mass determinations to be performed in a closed container that
contains only the →weighing pan (Fig. 99a). This makes it
possible to perform weighments in any (e.g. corrosive) medium, in vacuum, under pressure, or at high temperatures.
For this purpose, a controlled electromagnet is built into the
load hanger that keeps a permanent magnet in suspension
at a separation of approximately 1 cm.

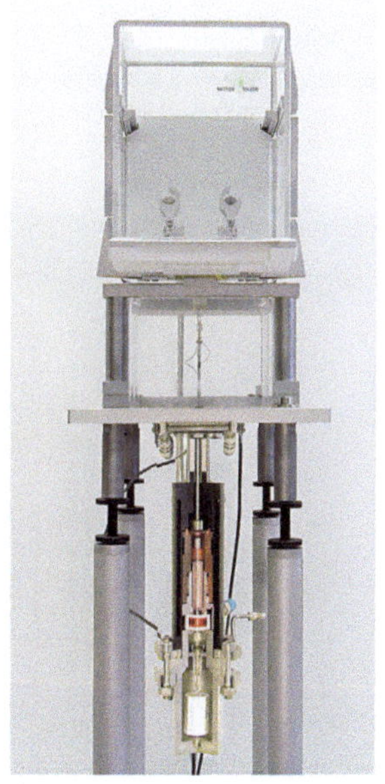

a)

Fig. 99a
Magnetic suspension balance
(Image by courtesy of Ruhr University
Bochum, DE)

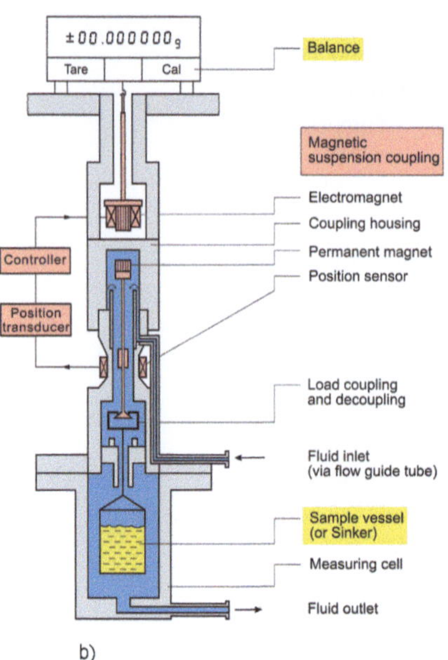

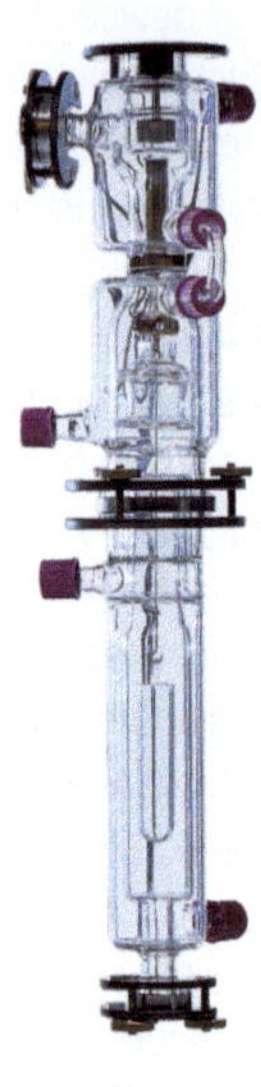

Fig. 99b
Diagrammatic cross section and
operating principle of a magnetic
suspension balance
(Image by courtesy of Ruhr University
Bochum, DE)

Fig. 99c
Detail view of a sample chamber
(Image by courtesy of Rubotherm
Präzisionsmesstechnik GmbH,
Bochum, DE)

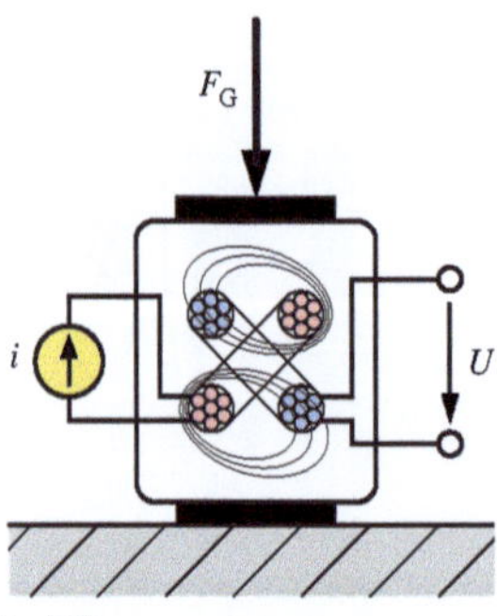

Fig. 100
A force sensor based on the magnetoelastic effect

F_G: weight force
i: exitation current
U: sensor voltage

magnetism

1. Physical phenomenon that produces forces between a
   source of magnetic fields (such as a permanent magnet)
   and other magnets, magnetically permeable objects or
   moving electrical charges (e.g. electric currents).
2. Physical property of certain materials that experience
   forces in a magnetic field. Two properties contribute to
   this behavior:
   1. Magnetic permeability: Magnetic permeability causes
      bodies to be attracted or repelled depending on their
      permeability to regions of inhomogeneous magnetic
      fields. Iron, nickel, and cobalt are highly permeable
      (ferromagnetic) and therefore strongly attracted by
      magnetic fields.
   2. Permanent magnetization: A body that is permanently
      magnetized is itself a cause of magnetic fields.

A weakly magnetizable or slightly permeable →weight piece
may experience spurious forces when it is magnetized or
exposed to a magnetic field.

magnetoelastic effect

Physical effect in which the magnetic permeability of a
ferromagnetic body is changed by elastic deformation
(Fig. 100). Materials that have this property include iron,
nickel, and cobalt.

main devices of the weighing instrument

Depending on the model, a →mechanical weighing instrument is composed of →load receptor (→parallel guide), load transmitter (→hanger, →lever system), and the →display device. In addition, an electromechanical weighing instrument has a measurement →converter, at least one →signal processing unit, and possibly a device for transmission of results (→data transmission).

main verification mark

In Germany, national →verification mark comprising
a) the →national verification mark and the →year mark for national verification (Fig. 101), or
b) the →national verification mark and the →year notation for national verification (Fig. 102).
→verification mark, →stamping mark

Fig. 101
Verification mark with year mark

Fig. 102
Verification mark with year notation

Maintenance Qualification

A term without official definition for a part of the →Equipment Qualification (EQ). Maintenance Qualification (MQ) comprises all measures necessary for planned maintenance, periodic calibration and, if necessary, adjustment, as well as cleaning of the equipment.

mass

Mass (m) is a fundamental property of matter, independent of location.
1. Physical quantity which can be ascribed to any material object and which gives a measure of its quantity of matter [OIML D 28].
2. Property of a body that results in its inertia to change in its state of motion, as well as its attraction to other bodies (→gravitation) ([DIN 1305] 2).
3. One of the seven base quantities of the →International System of Units (SI). The SI unit of mass (→unit of mass) is the →kilogram (kg).

The mass of a body is usually determined by →weighing. The embodiments of a unit of mass, and of its fractions and multiples, are usually called →mass standards, or in legal metrology, →weight pieces.

mass attraction

→gravitation

mass comparator

Balance of particularly high accuracy that is designed for →mass comparisons at a certain →nominal load (from

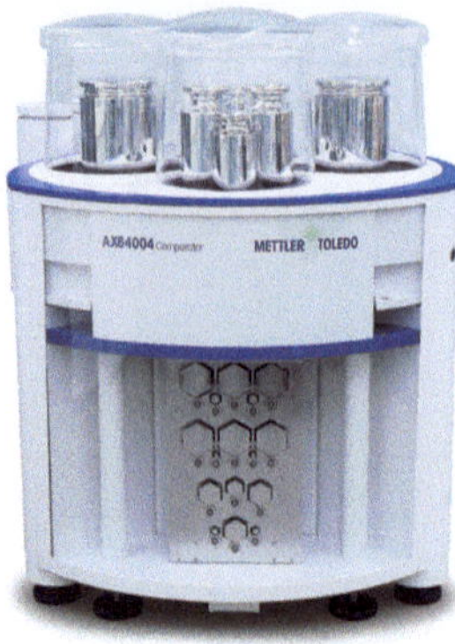

Fig. 103
Mass comparator with a capacity of
64 kg and a readability of 0.1 mg for
comparing mass standards

grams to metric tons, depending on the model) usually in a very small →weighing range (Fig. 103). Mass comparators generally have an electrical weighing range of up to 10^7 →scale intervals and indicate up to 10^9 scale intervals relative to the nominal load.

mass comparison

Determination of the difference between an unknown mass and a known mass (→reference mass). One reason why masses are compared is to reproduce the mass unit (→hierarchy of mass standards). Comparisons are ideally performed on balances specially designed for this purpose, so-called →mass comparators. →traceability

mass counter

A mass counter is a measuring instrument that determines the mass of a flowing liquid (mass flow) without the aid of other measuring instruments or data regarding the physical properties of the liquid. →Coriolis mass counter

mass flow

Mass flow measured as mass per unit of time. →mass counter

mass normal

→mass standard

mass standard

Embodiment of the →unit of mass (including its fractions and multiples) that is used to determine the →mass of other bodies. As opposed to →weight pieces, there are no special rules applying to mass standards. →reference mass

mass, conventional

→conventional mass

matrix code

→data matrix code

Max

Abbreviation for →'maximum capacity'.

maximum capacity

Upper limit *Max* of the →weighing range without consideration of an additional →maximum tare, i.e. the maximum capacity whose weight can be determined on a balance ([OIML R 76-1] T.3.1.1). →weighing capacity, →maximum safe load

maximum permissible deviation

→maximum permissible error

maximum permissible error

Largest deviation allowed (mpe) from a specified value
(→nominal value).
For →weight pieces, the maximum permissible error is
the amount of the difference between the actual mass or
→conventional mass of a weight piece and its nominal
value ([OIML R 111-1] 2.10) (→OIML weight classes).
According to European Directive →2009/23/EC, for non-
automatic weighing instruments the maximum permissible
error is the amount of the difference between the value indicated by the instrument and the value of the →test weight
([OIML R 76-1] T.5.5.4) (→maximum permissible error on
verification, →maximum permissible error in service).

maximum permissible error in service

The →maximum permissible error (positive or negative) of a
legally relevant measuring instrument in service (operation),
e.g. after →verification. For →non-automatic weighing instrument, the maximum permissible error in service is twice
the →maximum permissible error on verification (Tab. 3)
([OIML R 76-1] 3.5.2).

maximum permissible error on verification

→Maximum permissible error between the →measurement
value of a verified balance and the corresponding correct
true value determined with →standard weights at the time of
→verification ([OIML R 76-1] 3.5.1) (Tab. 3 and Fig. 104).
The error limit on verification apply for →initial verification
as well as →subsequent verification. For the →inspection,
the →maximum permissible error in service apply, which
are twice as large. →accuracy classes of weighing instruments, →accuracy classes of weight pieces

Permissible deviation (mpe)		Load in verification scale intervals (e)			
Max. permissible error on verification	Max. permissible error in service	I	II	III	IIII
0.5 e	1 e	0...50000	0...5000	0...500	0...50
1 e	2 e	50000...200000	5000...20000	500...2000	50...200
1.5 e	3 e	200000...	20000...100000	2000...10000	200...1000

Tab. 3
Maximum permissible errors on verification for weighing instruments according
to OIML R 76-1:
Maximum permissible error as a function
of load (both in →verification scale intervals e) ([OIML R 76-1] 3.5.1)

Fig. 104
Maximum permissible errors in service for weighing instruments according to OIML R 76-1

n: number of verification scale intervals
mpe: maximum permissible error
e: verification scale interval
I…IIII: accuracy classes of weighing instruments

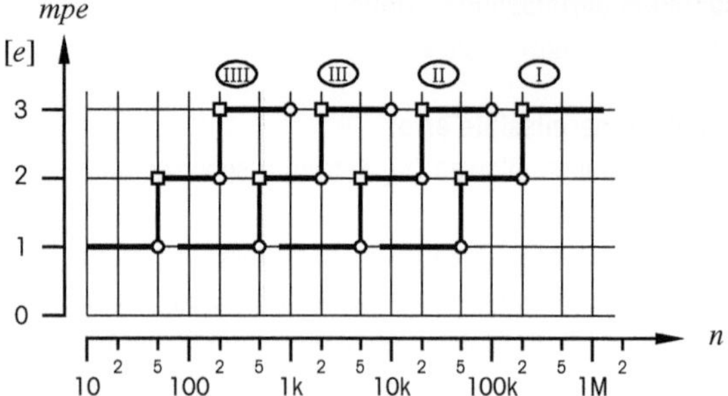

maximum permissible mass difference
→Maximum permissible error (mpe) of the mass of a →weight piece from its nominal mass.

maximum rated load
→maximum safe load

maximum safe load
Largest static load limit Lim that can be carried by a weighing instrument without permanently altering its metrological properties, and without the risk of damage, but which cannot usually be measured ([OIML R 76–1] T.3.1.7). It is always greater than or equal to the →maximum capacity plus the additive →maximum tare effect.

maximum tare
Largest applicable tare (→maximum tare effect).

maximum tare effect
Admissible →maximum capacity of the additive or subtractive →tare device with the denotation
additive maximum tare, e.g.   T = +20 kg
of a subtractive maximum tare, e.g.   T = –10 kg
([OIML R 76-1] T.3.1.6).

mean sample value
1. Arithmetic →mean value of a →sample.
2. A term used in →prepackage process control for the arithmetic →mean value (in units of mass or volume) of the →fill quantities of a sample.

mean value
Designation for a value $\bar{x}$ that is allocated applying a particular formula to n prescribed values $\{x_i\}$ and that lies between the largest and the smallest of these values. In con-

nection with weighing, the arithmetic mean value (arithmetic mean, average)

$$\bar{x} = \frac{1}{n}\sum_{i=1}^{n} x_i = \frac{1}{n}\left(x_1 + x_2 + \ldots + x_n\right)$$

is of significance.

mean value trace

Term used in →prepackage process control to designate the graphical representation of the mean sample values on the →control chart.

measurand

Physical quantity intended to be measured, e.g. mass, density, or volume ([VIM:2008] 2.3). →measurement value

measurement bridge

Electric circuit with several components for (accurate) measurement of electrical impedances (Fig. 186). →Wheatstone bridge

measurement cylinder

Cylindrical measuring vessel, usually of glass, plastic, or steel, with a volume scale, for measuring volumes of liquids (→volumetry), usually accurate to within 0.5% or less (Fig. 105).

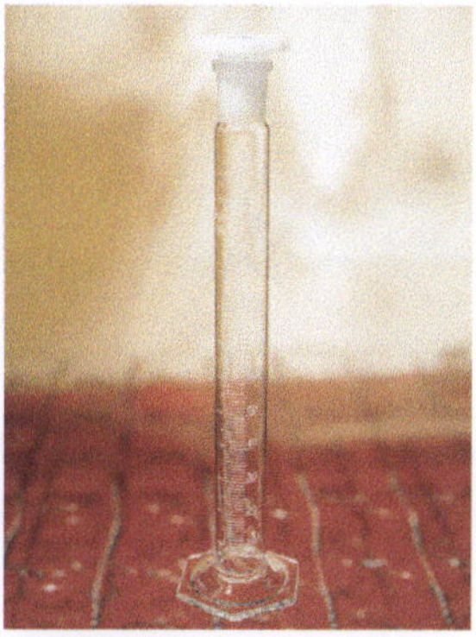

measurement deviation

1. Measured quantity value minus a reference quantity value; error. ([VIM:2008] 2.16)
2. →Measurement result minus the true value of the measurand. Since no true value can be determined, in practice an agreed true value is used. ([VIM:1993] 3.10)
3. Deviation from the true value of a value obtained by measurement and assigned to the measurand. ([DIN 1319-1] 3.5)

Fig. 105
Measurement cylinder
(Image by courtesy of DURAN Produktions GmbH & Co. KG, Mainz, DE)

measurement mark

Settable device (mark, pointer) that indicates a defined mass value of the weighing sample.

measurement pipette

→Pipette with a scale for measuring volumes.

measurement principle

Physical basis of a (quantitative) comparison that is to be performed between →measurand and →measurement unit, e.g. proportionality of →mass and →weight force as basis of a mass determination. →physical weighing principle

measurement result
Set of quantity values being attributed to a →measurand together with any other available relevant information.
A measurement result is generally expressed as a single measured quantity value (→measurement value) and a →measurement uncertainty. ([VIM:2008] 2.9) →measurement result of a weighing

measurement result of a weighing
Results of a weighing operation may be the →mass, the →weighing value, or the →conventional mass, where necessary taking into account corrections (e.g. →air buoyancy correction) and the →measurement uncertainty. With the aid of clearly defined relationships, the measurement result of a weighing can be used to derive →measurement values of various other →measurands such as, for instance, throughput (→mass flow, measurement value mass divided by time), or →density (measurement value mass divided by volume).

measurement signal
The physical or numerical quantity in the signal chain that is assigned to the measurand (→measuring chain).

measurement time
The time that elapses after placement of the load on the weighing instrument and the indication of the stable →measurement value. →settling time, →weighing time

measurement transducer
A device for converting an input measurand into an output measurand of a different physical type, also called a →converter. →load cell

measurement uncertainty
1. Non-negative parameter characterizing the dispersion of the quantity values being attributed to a measurand, based on the information used ([VIM:2008] 2.26).
2. A parameter which is assigned to the measurement result that describes the scatter of values that can reasonably be assigned to the measurand. ([VIM:1993] 3.9)
This parameter, i.e. the measurement uncertainty, is usually expressed as the →standard uncertainty u or the expanded measurement uncertainty U (→coverage interval). The mea-

surement uncertainty is obtained from a statistical analysis
of a series of observations (uncertainties of type A) and from
a non-statistical analysis or other information (uncertainties
of type B). Instructions for determining the measurement
uncertainty are contained in [GUM]. →uncertainty

measurement unit
Real scalar quantity, defined and adopted by convention,
with which any other quantity of the same kind can be compared to express the ratio of the two quantities as a number
([VIM:2008] 1.9).

measurement value
1. Quantity value attributed to the measurand. ([VIM:2008]
   2.10)
2. The value determined from the →indication of a balance
   as the product of a numerical value and a unit, e.g.
   mass m = 200 mg. The measurement value of a weighing is often also the →weighing result (→measurement
   result). →weighing value, →measurand

measurement value converter
A device that converts an analog →measurement value into
a digital measurement value (→analog-digital converter),
a digital measurement value into an analog measurement
value (→digital-analog converter) or characters of a coded
measurement value into characters of another code (code
converter).

measurement value deviation
→measurement deviation

measurement value drift
→drift

measuring chain
A sequence of series-connected elements, usually with a
measurement sensor as first element and a measurement
value emitter as last element. The →measurement signal
passes through the individual elements in the chain and,
while doing so, is converted (→converter) several times.

measuring container
Container with calibrated →volumes for measuring volumes
of liquid, e.g. →volumetric flask, →measurement cylinder,
→pipette, →burette, →pycnometer.

measuring instrument
Instrument that either alone or in conjunction with auxiliary devices is used to perform a measurement.

Measuring Instruments Directive
European Directive for a total of ten different instruments and system types with a measurement function (abbreviated "MID"). It contains all technical requirements and →error limits for the relevant types of measuring instruments and is applicable to all measuring instruments that perform measurement tasks in the public interest, in health care, public safety and security, environmental protection, consumer protection, tax collection, and general commercial applications. Among others, →automatic weighing instruments are governed by this directive.
The Weighing Instruments Directive is implemented as national law in the EEA and Switzerland as follows: If a country already has a national law relating to one of the instruments covered in the Measuring Instruments Directive, the corresponding module of the Measuring Instruments Directive must be implemented as national law and replace the formerly existing law. If a country has no national law relating to an instrument type, implementation of the corresponding module in national law is voluntary. →2004/22/EC

mechanical advantage
→ratio of mechanical advantage

mechanical weighing instrument
A balance that uses mechanical means of →load compensation. The weight as →measurand is compensated by mechanical means and represented by optical or other non-electrical aids (e.g. →substitution balance, →sliding weight balance, →spring scale). →design and function of a mechanical balance

medium accuracy
→Directive on Medium Accuracy Weights, →error limit class

Medium Accuracy Weights Directive
Medium Accuracy Weights Directive →71/317/EEC

METAS
METAS stands for the 'Swiss Federal Office of Metrology' (originally 'Metrologie und Akkreditierung Schweiz'), the Swiss →national metrology institute with headquarters at Bern-Wabern (www.metas.ch).

method
1. Means or activity with which to perform an operation
   (e.g. a measurement), or obtain a result (e.g. a →meas-
   urement result), in an orderly and systematic manner.
2. All settings of a measuring instrument (configuration)
   that are necessary for a measurement.
   2.1 In the case of a weighing instrument: Setting of
       parameters, for example →stand-still detector, filter
       parameters (→filter), operating mode (e.g. simple
       weighing, percent weighing, weighing method;
       →weighing instrument functions) →operating
       modes of a weighing instrument.
   2.2 In the case of a dryer: Setting of parameters, for
       example →drying program, drying temperature,
       →switchoff criterion, drying time, and target weight.

method parameter
→method

metric carat
Special unit of mass used for gemstones (diamonds)
(unit symbol "ct"). One metric carat is one fifth of a gram:
1 ct = 0.2 g. In legal use, the unit may only be used to
describe the mass of gemstones.

metric system of units
The metric system of units originally referred to a group of
units that were all derived from the meter. The units were for
commercially important physical dimensions of length, area,
volume, and mass (unit defined by mass of a specified volume of water).
In modern metrology, a system of units is no longer traceable to a single unit, but all units of the system are traceable to a certain small number of basic units. In this sense,
today's →International System of Units was developed from
the metric system. It is the modern form of the MKSA metric
system which has been expanded to seven basic units.

metric ton
The metric ton (unit symbol "t", sometimes referred to as
"tonne") is the one thousandfold multiple of the →kilogram:
1 t = 10^3 kg. →units of mass

metric unit
Unit that belongs to the →metric system of units (→Interna-
tional System of Units).

metrological characteristics of a weighing instrument

A weighing instrument should do the following:

a) indicate a sufficiently accurate →measurement value for the true value of the →mass of a →weighed object (→trueness);

b) for weighments performed under identical conditions, produce →weighing results that are as identical as possible (→repeatability);

c) respond with a →measurement value change to smallest possible changes in the load (→discrimination);

d) produce the correct change in →measurement value for a change in load (→sensitivity);

e) minimize the influence of external factors (i.e., temperature, humidity, etc.) on the measurement value (→environmental influence).

metrological test

1. Determination of the measuring characteristics (→specifications) of a weighing instrument by suitable testing procedures.

2. In legal metrology, this is part of an official test (e.g. →verification) of a measuring instrument to evaluate its metrological behavior. Generally this refers to the recording of error curves, determination of the →repeatability, etc. under various conditions (e.g. eccentric loading, different temperatures).

metrological testing of weighing instruments

The metrological testing of weighing instruments essentially comprises:

a) Testing of the →trueness at →minimum load, →maximum capacity, and various intermediate loads (recording of the error curve, →nonlinearity , →sensitivity);

b) Testing of the trueness in →eccentric loading;

c) Testing of the trueness in →tilt;

d) Testing of the →repeatability;

e) Testing of the →discrimination;

f) Testing for agreement of the display and printout device;

g) Testing of individual components (devices for leveling, zeroing, arrestment);

h) Testing of →auxiliary devices.

metrologically relevant

Subassemblies, →modules, parts, components, or functions of a →weighing instrument are considered metrologically relevant if they may influence the weighing result or any other →primary display ([OIML R 76-1] T.2.9).

metrology

The science of measurement.

metrology mark

Synonym for →'Green M'.

mg

Unit symbol for the mass unit →milligram.

µg

Unit symbol for the mass unit →microgram.

microbalance

→Analytical balance designed for microanalysis with a →weighing capacity of typically between 5 g and 20 g and a →readability of 1 µg (Fig. 106). →weighing instrument of special accuracy

microdispenser

Dispenser for filling small quantities. In combination with, for instance, a →precision or →analytical balance for weighing and filling small amounts of powder of approximately 100 mg or less, depending on the properties of the weighing sample.

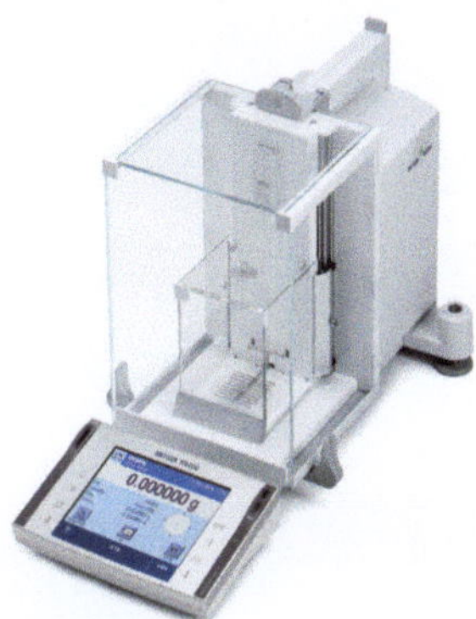

Fig. 106
Microbalance
Weighing capacity 50 g;
readability 1 µg

microgram

The microgram (unit symbol "µg") is the one-millionth part of the →gram and therefore the one-billionth part of the →kilogram: $1\ \mu g = 10^{-6}\ g = 10^{-9}\ kg$. →unit of mass

MID

Abbreviation for →'Measuring Instruments Directive'.

milligram

The milligram (unit symbol "mg") is the one-thousandth part of the gram and therefore the one-millionth part of the →kilogram: $1\ mg = 10^{-3}\ g = 10^{-6}\ kg$. →unit of mass

Min

Abbreviation for →'minimum capacity'.

minimum capacity

Lower limit *Min* of the →weighing range, below which the weighing results may be subject to an excessive relative error ([OIML R 76-1] T.3.1.2).

minimum load

1. Load below which the weighing results are subject to a high relative uncertainty. →accuracy classes of weighing instruments, →minimum sample weight
2. Load below which the weighing instrument no longer functions. This can, for example, be the case with →mass comparators whose →load range is restricted.

minimum sample weight

Smallest sample weight (→weighed-in quantity, →net weight) required for a weighment to just achieve a specified relative accuracy of weighing. Provided that →systematic errors have already been corrected, the minimum sample weight m_{min} can be determined from the allowed uncertainty U and the repeatability of the weighing s_{RP}

$$m_{min} = \frac{k}{U}\, s_{RP}$$

k →expansion factor

Corresponding requirements are described, for instance, in pharmacopeias (e.g. →USP $U = 0.1\%$, $k = 3$ [USP<41>] [12]), or may be defined in the user's process specifications.

minimum weight

Simplified term for →minimum sample weight.

minus deviation

Term used in →prepackage process control to designate (minus) errors (Tu limit) defined in tables in the →Prepackaged Products Directive. Twice this value is called the Tu2 limit, and any packages that fall below this limit must not be put into circulation. In addition, only 2% of all packages may lie between these two limits. →e-mark

MinWeigh®

→Application module that warns the user if the sample weight is less than the →minimum sample weight.

modular concept

A modular concept allows individual components (→modules) of a weighing instrument to be tested separately, if testing of the complete weighing instrument is not possible or if the weighing instrument should be composed of a combination of different modules. The modular concept requires

---

[12] At the time of printing, USP chapter 41 was in revision and the value proposed for k was 2.
® Registered trade mark of METTLER TOLEDO

the →error limit component p_i for a module that contributes
to the total measurement error to have a value between
0.3 and 0.8. Exceptions are an error limit component of
0 for modules whose operation is purely digital, and 1 for
→weigh modules that contain all metrological components
except a digital →display.

module
Part of a weighing instrument that can perform specific functions and be separately tested. Typical modules are →load
cell, →control unit, →weigh module, →terminal, →digital
display. ([OIML R 76-1] T.2.2)

Mohr´s balance
→Mohr-Westphal balance

Mohr-Westphal balance
An instrument for determining the density of liquids or solid
bodies that was developed by pharmacist Karl Friedrich
Mohr (*1806–†1879) and improved by Georg Westphal
(*1836–†1902). The instrument is used to measure the
hydrostatic →buoyancy by weighing the body whose density is being determined first in air and then in the liquid
(Fig. 107). →density determination, →hydrostatic balance

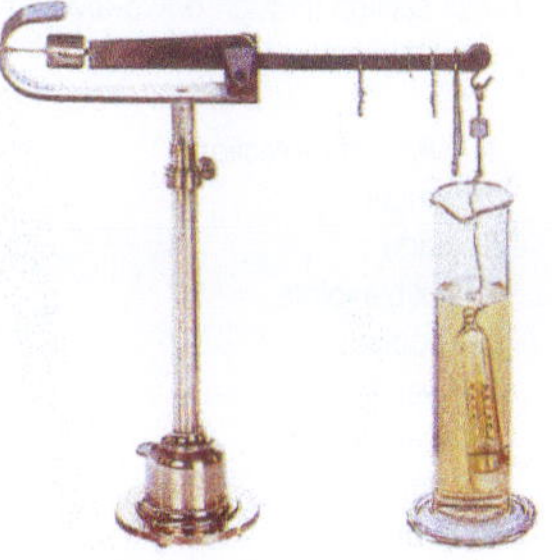

Fig. 107
Mohr-Westphal balance
(Image by courtesy of Gassner Wiege
und Messtechnik, Salzburg, AT)

moisture content
The relative mass content of the liquid in the total mass of
a mixture consisting of solids and liquids. For example, the
moisture content of grain is the difference in mass before
and after drying divided by the mass of the grain before drying. →dryer

momme
(Also "monme" or "monnme".) →Nonmetric unit of measure
for pearls, used in Japan (unit symbol "mo").
1 mo = 3.75 g.

Monobloc
Vendor-specific name of a technology for manufacturing
→monolithic load cells (Fig. 108).

monolithic load cell
Load cell in which the →parallel guide, one or more
→levers, the →hanger, all the →flexible bearings and
→force links, as well as the →overload protection, are
produced from one single piece (Fig. 108). By means of
suitable fabrication technologies, all of these mechanical

elements are formed from a single block of metal by separation (e.g. water-jet cutting or electrical discharge machining).

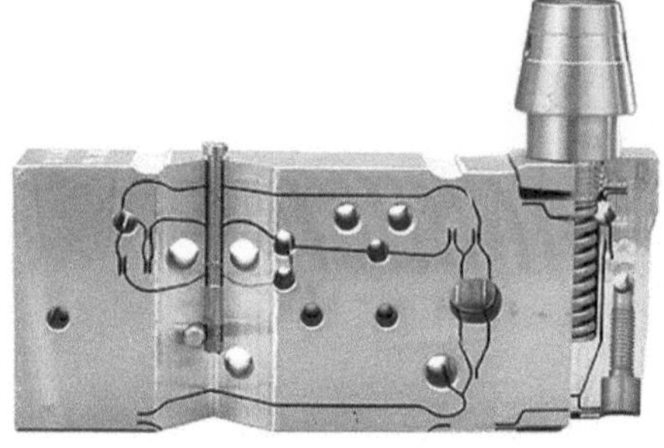

Fig. 108a
Monolithic EMFC load cell (without electrodynamic compensator). For greater clarity, the load cell is cut away in the left one-third and at the extreme right.

Fig. 108b
Cross section through a two-lever monolithic load cell

1: overload protection
2: hanger
3: guide
4: flexible joints
5: coupling 1
6: lever 1
7: lever bearing 1
8: coupling 2
9: lever 2
10: lever bearing 2
11: console

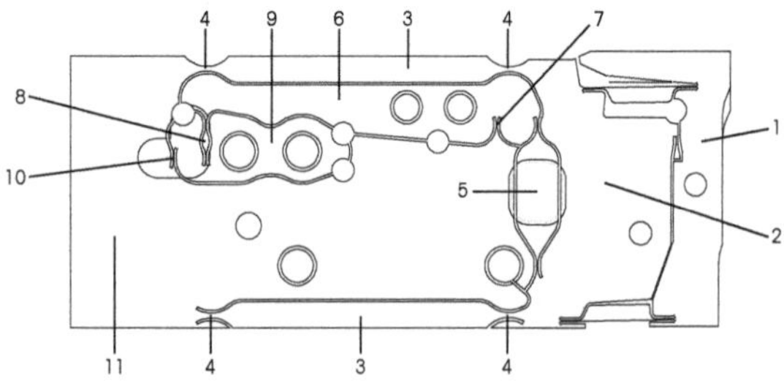

monorail scale

A scale that is built into an overhead rail system where part of the overhead rail serves as →load receptor; a.k.a. →overhead rail scale (Fig. 109).

movable scale

→Non-automatic weighing instrument that is built into or integrated in a vehicle or similar device, with which the scale can be moved on wheels without using any tools or other equipment. In the case of a built-in scale, a complete scale is mounted on a vehicle (e.g. →postal scale in a mail vehicle), whereas in an integrated scale, parts of the vehicle are used for the instrument (refuse scale in refuse truck, forklift truck with scale). ([OIML R 76-1] T.1.2.11)

mpe

→maximum permissible error

Multi Range (MR)

A weighing instrument that has two non-displaceable ranges that start at zero, both of which have a finer readability than the →normal range. →multi-range weighing instrument (compare: →DeltaRange)

Fig. 109
Monorail scale with a weighing capacity 300 kg

multi-interval instrument
Weighing instrument with one →weighing range which
is divided into partial weighing ranges each with different
→scale intervals, and in which the partial weighing range is
automatically determined according to the load applied, both
on increasing and decreasing loads (Fig. 110b) ([OIML
R 76–1] T.3.2.6). →DeltaRange (compare: →single-range
balance, →multi-range weighing instrument)

multi-pan balance
Balance that has more than one pan (→single-pan balance), for instance the →Béranger scale or →three-knife
balance.

multi-range weighing instrument
Term used in non-technical language for weighing instruments to with at least two non-displaceable →fine ranges.
→multi-range weighing instrument

multicomponent weighing instrument
→Automatic gravimetric filling instrument for apportioning
(→apportion) or →weighing, with which preset weight values of different components can be supplied to, for instance,
a mixer by multiple apportionments or by weighing and
emptying.

multihead weigher
→combination scale

multiple interval
→multi-interval instrument (Compare: →multi-range weighing instrument)

multiple range
→multi-range weighing instrument (Compare: →multi interval instrument)

multiple range instrument
Weighing instrument with two or more →weighing ranges
with different →maximum capacities and →scale intervals
for the same →load receptor, each range extending from
zero to its respective →maximum capacity (Fig. 110a)
([OIML R 76–1] T.3.2.7). The weighing ranges can be assigned to different →accuracy classes. →Dual Range (compare: →single-range balance, →multi-interval instrument,
→DeltaRange)

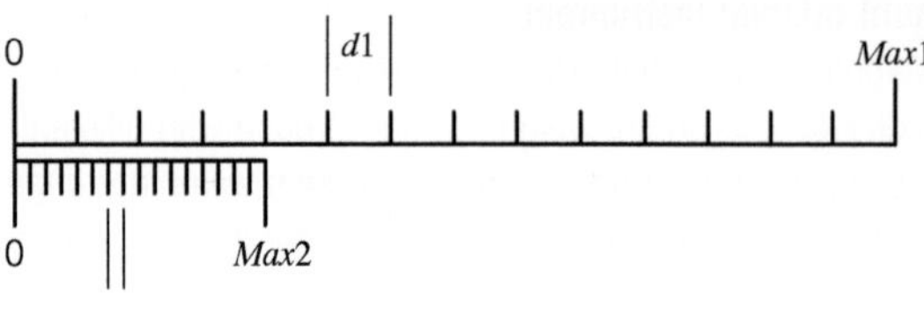
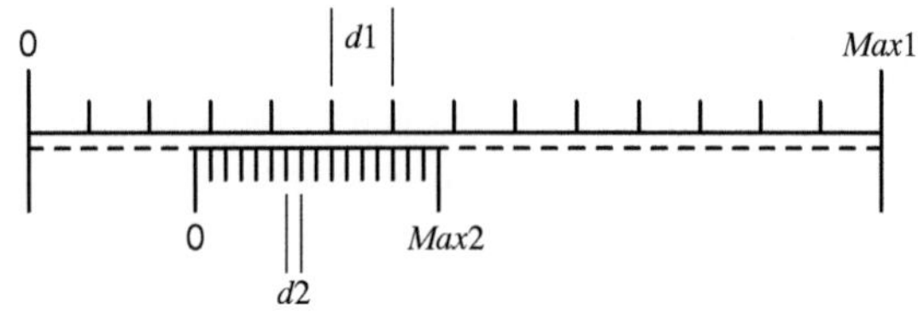

Fig. 110
Weighing ranges of a
a) multi-range instrument;
b) multi-interval instrument

multiuser system

A computer designed for use as a multiuser system allows the connection of multiple terminals (or networked weighing instruments) thereby allowing multiple users to work virtually simultaneously. Multiuser systems have a particular type of (operating) software that controls the data transfer to and from the mass storage devices connected to the computer, blocks certain areas of the working storage, assigns individual peripheral devices to the users, controls execution of the individual programs, etc.

multiuser weighing system

→multiuser system

n

→number of verification scale intervals, →number of scale intervals

N

1. Unit symbol for the unit of force →newton.
2. Symbol for →net value.

National Conference on Weights and Measures
An organization in the United States with more than 2400 members. The National Conference on Weights and Measures (NCWM) ensures uniformity, consistency and fairness in the marketplace. Weights and Measures regulatory professionals set standards and enforce uniform procedures to verify weight, volume, length or count, ensuring that consumers get the quantity that they pay for, and that businesses sell the quantity that they intend and advertise. (www.ncwm.net)

National Institute of Standards and Technology
→National metrology institute of the United States of America, with laboratories in Gaithersburg (Maryland), and Boulder (Colorado). (www.nist.gov)

national metrology institutes
The main responsibilities of the national metrology institutes are:
– Research and development relating to the base units;
– Safekeeping and development of the national standards;
– Implementation and communication of the internationally agreed base units;
– Test, approval, and conformity examination of measuring instruments used in legal metrology;
– Participation in national and international committees for metrology and Weights and Measures authorities.
A random selection of national metrology institutes is:
→NIST (USA), NRC-INMS (Canada), NIM (China), NMIJ (Japan), KRISS (Republic of Korea), NMIA (Australia), →PTB (Germany), →NPL (UK), →LNE (France), VSL (Netherlands), →METAS (Switzerland).

National Type Evaluation Program
→Type examination applicable in the USA for weighing devices used in legal-for-trade applications (Fig. 132). The National Type Evaluation Program (NTEP) is a program of cooperation between the →National Conference on

Weights and Measures (NCWM), the →National Institute of Standards and Technology (NIST), the states, and the private sector. NTEP evaluates the performance, operating characteristics, features and options of weighing instruments and measuring devices against the applicable standards. Following successful completion of the evaluation and testing of a device, an NTEP Certificate of Conformance is issued. (www.ncwm.net/ntep)

Fig. 111
National verification mark for verification in Germany

national verification mark

The national verification mark for Germany consists of a banner bearing the letter D, the →ordinal number of the respective →Weights and Measures authority, and a six pointed star (Fig. 111). Instead of the star, the respective ordinal number of the verifying →Weights and Measures office can also be used. →main verification mark

NAWI

Abbreviation for →'non-automatic weighing instrument'.

NCWM

Abbreviation for →'National Conference on Weights and Measures'.

net value

Indication of the weight value (→weighing value) of a load placed on a weighing instrument, after operation of a →tare device, often designated with symbol N ([OIML R 76-1] T.5.2.2). →net weight

net weight

Weight of a →weighed object after deduction of the weight of its packaging or of a transport device (→tare weight) with which it was weighed. →weighed-in quantity, →gross weight

newton

Unit of →force in the International System of Units (SI), named after Isaac Newton (*1643–†1727), unit symbol "N", 1 N = 1 kg·m/s^2. A force of 1 N imparts to a body of 1 kg an acceleration of 1 m/s^2.

NIST

Abbreviation for →'National Institute of Standards and Technology'.

noise
In physics, noise is an interference quantity that usually has
a wide frequency spectrum. Most measurement signals are
overlaid with noise, which limits the measurement →resolu-
tion. As far as possible, noise is eliminated from the measurement signal with the aid of →filters.
In association with weighing instruments, the most important sources of noise are as follows:
1. Electric noise from electronic components: Occurs in
   every electronic (measurement) circuit (e.g. →reference
   voltage of an →A/D converter).
2. Movement noise of the foundations at the →place of
   installation: Microseismic accelerations (caused by
   weather, ocean waves, and/or human activity, e.g. traffic, machines, manufacturing facilities, etc.) overlay the
   weighing signal. →vibrations

nominal capacity
→nominal load

nominal fill quantity
Term used in →prepackage process control to designate the
quantity of filling material indicated on the package. →fill
quantity

nominal load
→Nominal value of the →load range of a →weighing
instrument. →weighing capacity, →maximum capacity

nominal load range
→nominal load

nominal range
→Nominal value of the →weighing range.

nominal value
Approximate quantitative value, as a round number or as
a rounded value of another quantity. The nominal value
is generally a number with few significant digits, e.g.
→nominal load 200 g (for a balance with a →maximum
capacity of 215 g) or 1 kg for a weight piece of mass
1.00036 kg.

non-automatic weighing instrument
Weighing instrument that requires the intervention of an
operator during the weighing process to decide whether the
→weighing result is acceptable ([OIML R 76-1] T.1.2).

The decision as to whether a weighing result is acceptable covers all intelligent actions of the operator that affect the weighing result. This can be an action when the indication is stable (e.g. to trigger a →printout, to →tare, or to →set to zero the weighing instrument) or a possible adjustment of the weight of the weighing sample while observing the read-out. →Directive on Non-Automatic Weighing Instruments

non-interacting data output

On weighing instruments with data output: Connection point for digital or analog forwarding of the measurement value (e.g. for →auxiliary displays, →printer devices, EDP systems). "Non-interacting" mainly means that the measurement value is not falsified even in the case of extreme load conditions on the output caused by, for example, short or open circuit. In many cases, there is also a →galvanic separation provided. →protected interface

non-self-equilibrating instrument

→non-self-indicating instrument

non-self-indicating instrument

Weighing instrument in which the position of equilibrium is obtained entirely by the operator ([OIML R 76-1] T.1.2.5). →self-indicating instrument, →semi-self-indicating instrument

nonlinearity

1. Deviation of the →characteristic curve from the straight line between →zero load and →nominal load that is defined by the →sensitivity. By definition, the linearity deviation of the starting and finishing point of this straight line is zero, and a possible deviation of the sensitivity (slope of the straight line) does not count as linearity deviation.
2. Specification: Magnitude of deviation of the →characteristic curve from the straight line between →zero load and →nominal load, generally for increasing load (Fig. 112), usually expressed as a limit value in mass units, e.g. [g].

nonlinearity, differential

→differential nonlinearity

nonmetric mass unit

→Nonmetric unit of mass [13]. The most important group of nonmetric units are the →units of mass that are widespread in Anglo-Saxon countries. They are based on the grain:

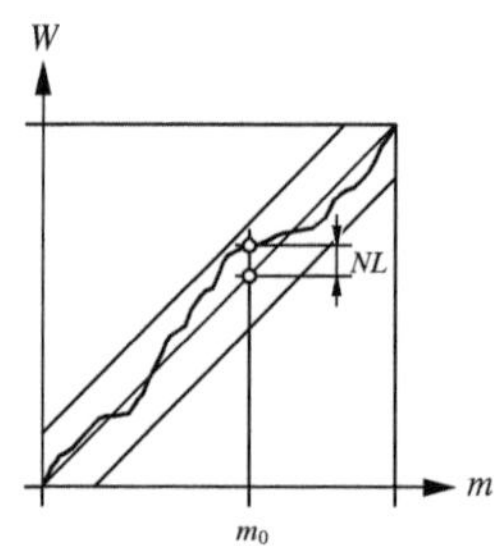

Fig. 112
Linearity deviation (exaggerated) between load m and weighing value W. A possible deviation of the sensitivity (slope of the straight line) does not count as linearity deviation.

Grain	1 GN = 64.79891 mg (exactly)

Avoirdupois units

pound	1 lb(lbm) = 7000 GN ≈ 453.6 g
ounce	1 oz = (1/16) lb ≈ 28.35 g
short ton	1 short ton = 2000 lb ≈ 907 kg
long ton	1 long ton = 2240 lb ≈ 1016 kg

Troy units

pennyweight	1 dwt = 24 GN ≈ 1.555 g
ounce	1 ozt = 20 dwt ≈ 31.10 g
pound	1 lbt = 12 ozt ≈ 373.2 g

In addition, there are many units of mass that are used for special applications, such as

Gemstones

→metric carat	1 ct = 0.2 g

Precious metals

→tael, Hong Kong	1 tl ≈ 37.429 g
tael, Singapore	1 tl ≈ 37.79936 g
tael, Taiwan	1 tl = 37.5 g

Pearls

→momme	1 mo = 3.75 g
kann	1 ka = 1000 mo = 3750 g

nonmetric system of units

System of units that deviates significantly from the metric
→International System of Units. An example is the system
of U.S. Customary Units in which, for instance, the units of
length are the mile (mi), yard (yd), foot (ft), and inch (in),
where

1 mi = 1760 yd

1 yd = 3 ft

1 ft = 12 in

→nonmetric unit of mass

nonmetric unit

→Measurement unit of a →nonmetric system of units.

normal distribution

Probability distribution, a.k.a. Gaussian distribution, whose
probability of occurrence decreases monotonically on
both sides of the expected value μ (so-called bell curve,
Fig. 113). The width of the probability distribution depends
on the →standard deviation σ. The expected value (→mean
value) and standard deviation completely describe the dis-

---

[13] Comprehensive tables of conversion factors between metric and nonmetric
units are contained in [NIST HB 44] and [Wildi].

tribution. Approximately 68% of occurrences lie within plus/minus one standard deviation (→expansion factor) about the mean value, approximately 95% within two standard deviations, and approximately 99.7% within three standard deviations (→coverage interval).

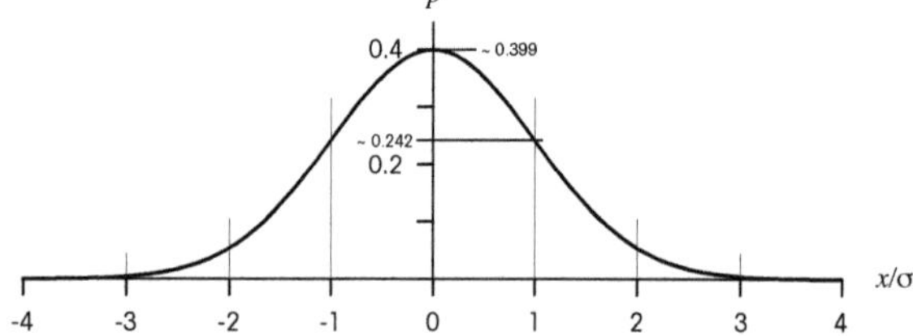

Fig. 113
Probability density function p of a normally distributed random quantity x with expected value $\mu = 0$ and standard deviation $\sigma = 1$

normal range
Weighing range of a →dual-range weighing instrument or →multi-range weighing instrument usually with greater →readability; also →coarse range. (Compare: →fine range)

Notified Body
Neutral and independent body that is nominated by the member states of the EU and according to European Directives is called upon in most conformity evaluation processes to perform, for example, →EC type examinations, EC verifications, etc. For instance, as part of the →type examination, a Notified Body tests and certifies that a →non-automatic weighing instrument that is representative of what is planned to be produced conforms to the stipulations of the →Directive on Non-Automatic Weighing Instruments →2009/23/EC. A list of notified bodies is available at www.welmec.org.

NPL
Abbreviation for 'National Physical Laboratory', the British →national metrology institute with headquarters at Teddington, Middlesex (www.npl.co.uk).

NTEP
Abbreviation for →'National Type Evaluation Program'.

number of scale intervals
Quotient of →maximum capacity *Max* and →division d

$$n_\mathrm{d} = \frac{Max}{d}.$$

→relative resolution

number of verification scale intervals
Quotient of →maximum capacity *Max* and →verification
scale interval *e*

$$n = \frac{Max}{e}\ .$$

([OIML R 76-1] 3.2.5)

numerical interval
Difference between two successive numbers on a numerical
→scale. →readability, →scale interval

obligation to record

The following obligations to record apply in relation to:

1. Good Laboratory Practice:
   Companies that operate under →Good Laboratory
   Practice (GLP) have an obligation to archive quality-
   relevant records of laboratory tests and to preserve them
   for a specified period of time which may, however, vary
   from country to country. Examples of records under GLP
   are test plans, raw data, final reports, and inspection reports. The details are laid down in the OECD Principles of
   Good Laboratory Practice, which have been implemented
   as national law in the EEA and Switzerland in the form of
   European Directive →2004/10/EC.

2. European Directives:
   Many European Directives, including those regarding
   the technical harmonization of products (→Directive
   on Non-Automatic Weighing Instruments, →Measuring
   Instruments Directive, →Low Voltage Directive, →EMC
   Directive, →ATEX 95 Directive, etc.), define the requirements for the recording obligations to be fulfilled by the
   manufacturers of the respective instruments. These relate
   to manufacturing drawings, internal and external test
   reports, design calculations, etc. In many cases, an obligation to record is stipulated for ten years after the manufacture or bringing into circulation of the last product.

3. Prepackaged Products Directive:
   Producers of →prepackages are obliged to keep records
   regarding compliance with the fill quantity requirements
   (mean value, minus deviations, spreads, date of the
   inspection) and present them to the inspecting authorities
   if requested to do so. The necessary control measuring
   instruments and procedures, as well as procedures for
   verification of →fill quantities by the responsible authorities, are also described in the →Prepackaged Products
   Directive. In practice, in a sample inspection the following data are usually recorded: Date, name of the tester,
   date and time of the test, designation of the filling plant,
   product designation, →sample size, number of tolerance
   violations, →nominal fill quantity, →target fill quantity,
   →mean sample value, →repeatability, mean →tare
   weight, possibly tare dispersion. This regulation applies
   in the EEA and Switzerland.

Oechsle hydrometer

→Hydrometer for measuring the sugar content of a liquid
developed by Christian Ferdinand Oechsle (*1774, †1852).

OIML

Abbreviation for 'Organisation Internationale de Métrologie Légale' (French for →'International Organization for Legal Metrology'). →BIML

OIML certification system for measuring instruments

A voluntary system for the issue, registration, and use of certificates regarding the conformity of →measuring instrument types with relevant →OIML recommendations and documents. These recommendations must satisfy certain demands with regard to their technical content.

The International Bureau for Legal Metrology (→BIML) carries a list of the measuring instruments for which suitable OIML recommendations and documents exist (www.oiml. org). OIML certificates of conformity are issued by notified issuing bodies of the OIML member countries on the basis of a conformity test in a competent laboratory.

OIML recommendations and documents

International recommendations for the metrological and technical characteristics of measuring instruments and their verification procedures that are published by →OIML. According to the OIML treaty, the member states are obliged to integrate the recommendations as far as possible into their national regulations. For gravimetric and volumetric determinations, the following documents are relevant:

Load cells:	OIML R 60
Weighing instruments:	OIML R 50, R 51, R 61, R 76, R 87, R 106, R 107, R 134
Weight pieces:	OIML R 52, R 111, D 28
Volumetric flasks:	OIML R 4

(See literature references)

OIML weighing instrument classes

→accuracy classes of weighing instruments

OIML weight classes

Classification of →weight pieces according to error limits that are defined in OIML R 111-1 "Weights of classes E1, E2, F1, F2, M1, M1–2, M2, M2–3 and M3". This recommendation defines the denominations ($1\times$, $2\times$, 5×10^n g) and characteristics for weights from 1 mg to 5 t in the nine defined classes. The maximum permissible relative error (→maximum permissible error, mpe) for weights of class E1 is 0.5×10^{-6} (for weights ≥ 100 g) and increases per class by a factor of approximately 3, and per two classes by a factor of 10, to 0.05% for class M3 [14]. The shape of the weights (Fig. 114), the materials to be used and their densities, the surface qualities, the magnetic characteristics ($\rightarrow$magnetism), etc. are specified for each class. The calibration uncertainty U must not exceed 1/3 of the mpe at $k = 2$, which corresponds to a $\rightarrow$standard uncertainty u of 1/6 mpe. The deviation of the $\rightarrow$conventional mass from the $\rightarrow$nominal value must not occupy more than the remainder of the mpe. $\rightarrow$calibration

OIML class	mpe (m >100 g)	u
		$\leq {}^1/_6 \cdot mpe$
E1	0.00005%	0.000008%
E2	0.00016%	0.00003%
F1	0.0005%	0.00008%
F2	0.0016%	0.0003%
M1	0.005%	0.0008%
M1-2	0.010%	0.0017%
M2	0.016%	0.0027%
M2-3	0.03%	0.005%
M3	0.05%	0.008%

Tab. 4
OIML weight classes
Maximum permissible relative deviation (mpe) and maximum permissible standard uncertainty on calibration (u) as a function of the weight class (Note: Weights with denominations less than 100 g allow higher relative deviations.)

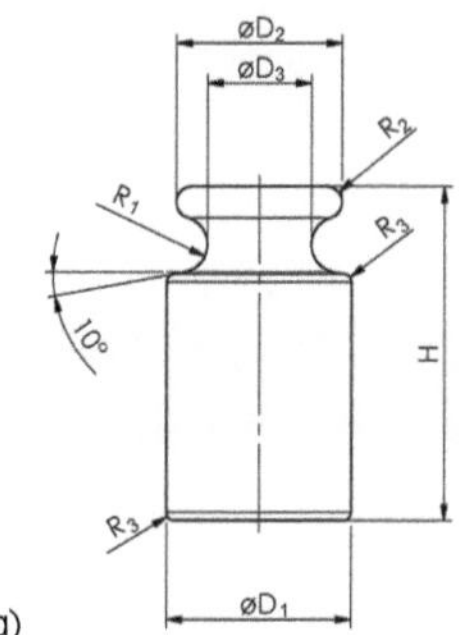

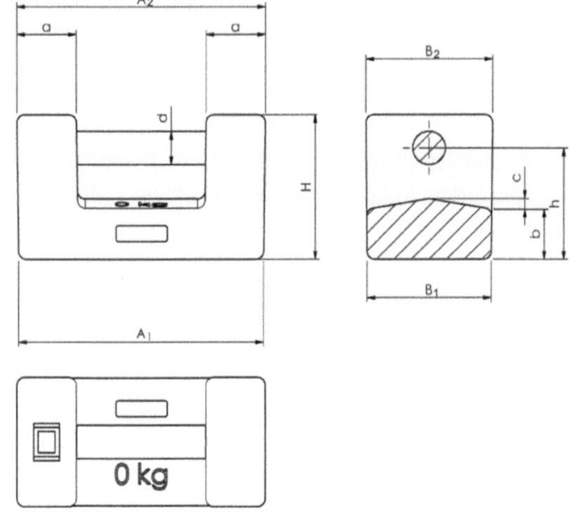

Fig. 114
Examples of shapes of OIML weights
a) cylindrical weight;
b) bar weight

(Images by courtesy of BIML, Paris, France)

---

[14] except for classes M1-2 and M2-3

onboard truck scale
→vehicle on-board weighing system

operating modes of a weighing instrument
A weighing instrument can have various different switchable operating modes that are individually designated by corresponding symbols, characters, or text (OIML R 76-1, 4.20). Examples:
> Weighing mode: Different weighing ranges, load-carrier combinations, tare inputs, etc.
> Weighing mode switched off: Counting scale, percentage weighing, indication of calculated values, totals, statistics, etc. (→application module).

operating principle of a mechanical balance
→design and function of a mechanical balance

operating principle of an electromechanical balance
→design and function of an electromechanical weighing instrument

operating principle of an electronic balance
→design and function of an electromechanical weighing instrument

operating temperature range of a weighing instrument
Ambient temperature at which the weighing instrument can be used. →temperature range

Operational Qualification
Part of →Equipment Qualification (EQ). The Operational Qualification (OQ) documents that the instrument functions according to the defined specifications in the intended environment.

operator guidance
Information given to the user on the display (usually a →display screen) regarding which operations must be performed in a particular case.

ordinal number
An identification number of a German →Weights and Measures authority. Table 5 contains a list of these ordinal numbers. →verification, →main verification mark, →repairer identification mark

output signal
Usually electonic signal on the output of a module, e.g. a measurement amplifier, an analog load cell or an electronic weighing instrument.

State	Weights and Measures authority	Ordinal number	Identifica-tion letter	Identification number as Notified Body
Baden-Württemberg	Mess- und Eichwesen in Baden-Württemberg, Reg.Präs. Tübingen	22	A	103
Bayern	Bayerisches Landesamt für Mass und Gewicht	23	B	104
Berlin/Brandenburg	Landesamt für das Mess- und Eichwesen Berlin-Brandenburg	1	C	106
Bremen	Der Senator für Arbeit – Landeseichdirektion – Bremen	19	D	107
Hamburg	Eichdirektion Nord	7	E	108
Schleswig-Holstein	Eichdirektion Nord	7	M	—
Hessen	Hessische Eichdirektion	10	F	109
Mecklenburg-Vorpommern	Wirtschaftsministerium Mecklenburg-Vorpommern, Landeseichbehörde	14	P	110
Niedersachsen	Mess- und Eichwesen Niedersachsen (MEN)	8	G	111
Nordrhein-Westfalen	Landesbetrieb Mess- und Eichwesen Nordrhein-Westfalen	11	H	112
Rheinland-Pfalz	Landesamt für Mess- und Eichwesen Rheinland-Pfalz	4	K	113
Saarland	Landesamt für Umwelt- und Arbeitsschutz – Eichaufsichtsbehörde Saarland	13	L	114
Sachsen	Sächsisches Landesamt für Mess- und Eichwesen	12	R	115
Sachsen-Anhalt	Landeseichamt Sachsen-Anhalt	6	S	116
Thüringen	Landesamt für Mess- und Eichwesen Thüringen	15	T	118

Tab. 5
Ordinal numbers, identification letters of the German Weights and Measures authorities, and their identification numbers as notified bodies.

over/under scale

Self-equilibrating or semi-self-equilibrating comparison weighing instrument that indicates the deviation of the mass of a sample from a set target mass as an excess or a short-fall.

overhead rail scale

→monorail scale

overload indicator

A device that indicates overloading or underloading of the weighing instrument. In →electronic weighing instruments, the indication is shown in the →display.

overload lock
A locking device that prevents →weighing above the →maximum capacity (→weighing range) and protects individual components against overloading. →maximum safe load

overload protection
Mechanical device that protects the weighing instrument or →load cell from overloading by interrupting the transmission of force from the →load receptor to the load cell as soon as an excessive load or force acts on the load receptor. →overload lock

package

Goods filled into similar packaging containers in series operation, including the packaging container. The →fill quantities of packages that are marked by weight are predominantly →apportioned by →automatic weighing instruments or filling machines and tested by →checkweighers (usually by random sampling). Packages can also be monitored by bulk density in the plant by checkweighers when the weight of the packaging container is sufficiently constant or can be taken into account separately. →filling process control

packaging

(Also packing means). Product made of packaging material whose purpose is to wrap or hold together the packaged goods so that they can be shipped, stored, and sold. A package in the context of prepackage process control has a protective function, an application function, and an information function (→fill quantity and details of manufacturer).

pallet scale

→Scale for weighing pallets along with their contents (Fig. 115). →forklift scale

Fig. 115
Pallet scale (weighing capacity 2 t; readability 0.5…1 kg)

pan

1. On a small balance, the →load receptor that serves directly to accommodate the →load (→load pan) or →weight pieces (→weighing pan), a.k.a. platter. (Compare: →platform)
2. In →knife-edge bearings, the mechanical part that makes contact with the →knife-edge (Fig. 90), a.k.a. flat or bearing. It is made of hardened steel, synthetic sapphire, or any other material that is similar in hardness and strength to the knife edges with a flat surface (also called plan bearings), V-shaped notches (V bearings), or shaped in the form of a ring (ring bearings).

pan brake

A device for slowing and ultimately stopping the oscillations of the →hanger and →pan of a mechanical →analytical balance. →design and function of a mechanical balance, →low-level pan

parallel guide

Mechanical arrangement for guiding the →load receptor in the form of a →parallelogram (Fig. 116). The parallel guidance system compensates the mechanical torque that arises on eccentric loading of the load receptor (→eccentric

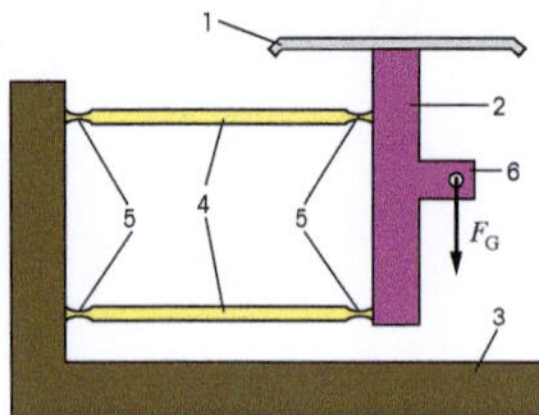

Fig. 116
Parallelogram

1: load receptor
2: hanger
3: base
4: guide
5: joint
6: force link to measurement
     converter
F_G: weight force

load) and thereby prevents the torque from affecting the result of the weighing. The two horizontal sides are formed by parallel →guides, one vertical side is connected to the base of the →weighing instrument, the other forms the hanger and is connected to the load receptor. At all four corner points there are →joints, which on →low-resolution instruments take the form of →pivot joints and on →high-resolution instruments the form of →flexible joints. The ideal parallelogram allows the →hanger and the attached load receptor only one degree of freedom: translation in the vertical direction. The remaining horizontal translations, as well as all rotations, are guided, with the result that the load receptor of →top-loading instruments is prevented from tipping. →Roberval scale

parallelogram

1. Geometry: Convex rectangle whose respective opposite sides are parallel.
2. Weighing instrument construction: →Parallel guide, first realized by Gilles Personne de Roberval (→Roberval scale).

parcel scale

→Scale for →weighing parcels, for instance in post offices (→letter scale) or in goods shipping.

parts counting

→piece counting system

passthrough sale

A sale in which a customer is served at multiple interconnected scales in a public point of sale (scale system). Allocation of the goods to the customer takes place via the salesperson allocation.

patient scale

→bed scale

pattern approval marks

(Fig. 117)
1. Symbol for the national type approval:
1.1 Measuring instruments (Fig. 117a)
     Top field: Four-digit number to identify the measuring instrument type, starting with 9 for non-automatic weighing instruments (national type approval for non-automatic weighing instruments only until 1992), starting with 10 for automatic weighing instruments

Bottom field: Four-digit sequential type number in which
the first two digits identify the year of approval (e.g. 89
for 1989).

1.2 Peripheral devices (Fig. 117b)
Peripheral devices receive the same approval symbols
as measuring instruments. Until 1988, however, the
following approval symbol was allocated to peripheral
devices:
Top field: Four-digit number to identify the type of
peripheral device
Bottom field: Four-digit sequential type number.

2.   EEC type approval mark
(as from 1 January, 1993, these symbols are no longer used for →non-automatic weighing instruments):

2.1 EEC type approval (Fig. 117c)
Top field: e.g. D for Germany, if the type approval was
granted by PTB, followed by the last two digits of the
year of approval
Bottom field: Four-digit number to identify the type of
measuring instrument, starting with 9 for non-automatic
weighing instruments, starting with 10 for automatic
weighing instruments.

2.2 Restricted EEC type approval (Fig. 117d)
P: no special descriptive-mark
E: descriptive-mark see 2.1 above.

2.3 General approval for EEC verification (Fig. 117e)
Top field: e.g. D for measuring instruments manufactured in Germany, followed by the last two digits of the
year of manufacture
Bottom field: Empty.

3.   Symbols for →EC type approval (→non-automatic
weighing instruments)
Not a special symbol but only an approval number that
in Germany (Notified Body: PTB) takes the following
form: DYY-09-XXX.
DYY: D for Germany
YY for the year in which approval was granted
09:    Annex of the Verification Ordinance [VO]
XXX: Serial number in the year of approval
(The nomenclature for the approval numbers of other
European notified bodies is contained in the WELMEC 2
guideline.)

pattern examination
Part of the verification procedure which serves to determine
whether

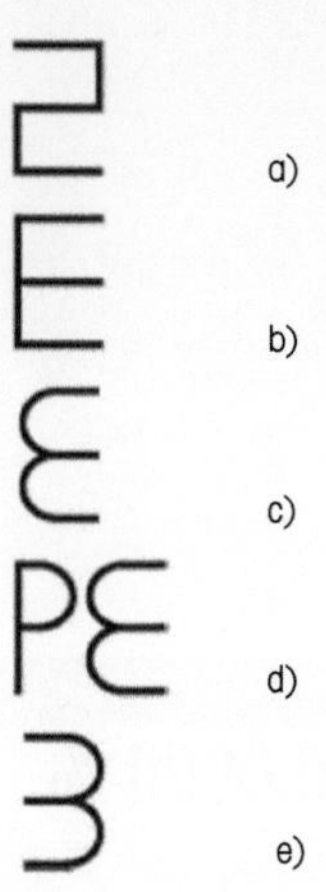

Fig. 117
Pattern approval marks up to
November 1988
a) measuring instruments;
b) peripheral devices up to
November 1988;
c) EEC type approval;
d) restricted EEC type approval;
e) general approval for initial EEC
verification

a) the type or design of the weighing instrument is admissible for verification;
b) the construction of the weighing instrument meets the requirements for verification;
c) the required destinations, inscriptions, and stamp emplacements are present.

PC, certified
→certified computer

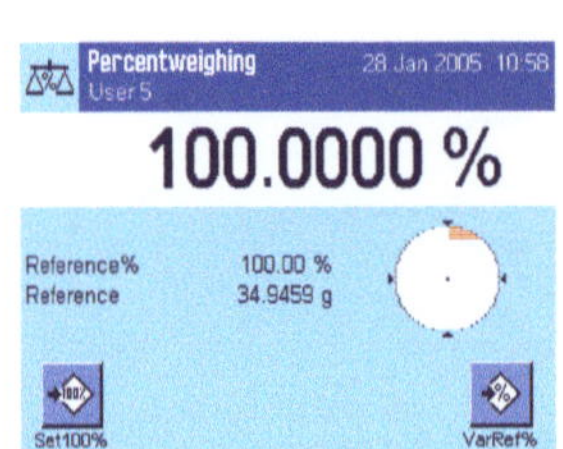

Fig. 118
Display of a percentage balance

percentage balance
A weighing instrument equipped with a scale or indicating device subdivided into percentages, e.g. balances that are used to determine the contents of water, starch, or impurities (Fig. 118). →dryer

Performance Qualification
Part of →Equipment Qualification (EQ). The Performance Qualification (OQ) documents that the instrument conforms to the requirements and specifications in routine operation.

period of verification validity
→validity period of verification

peripheral device
1. In general, peripheral devices are instruments and equipment that are connected to a computer by means of a data →interface (e.g. printer, screen, plotter connected to a PC).
2 In relation to a weighing instrument, a peripheral device is an additional device that repeats or further processes the weighing result. Examples: Printer, secondary display, keyboard, terminal, data storage device, code reader, personal computer) ([OIML R 76-1] T.2.3.5).

person scale
Low-accuracy →scale in private households for weighing people, usually executed as a →spring scale of →accuracy class ⎍ (Fig. 119).

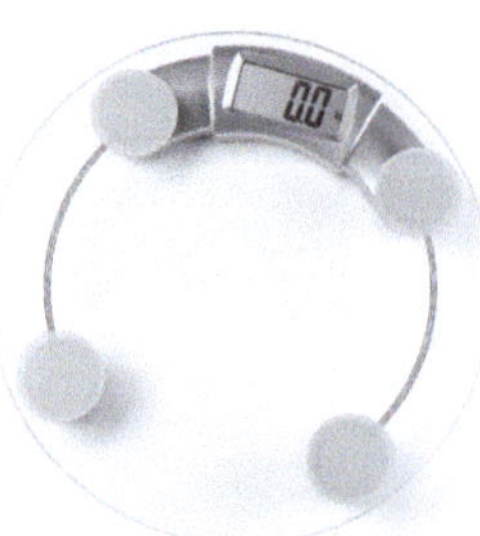

Fig. 119
Person scale
(Image by courtesy of
Trisa Elektro AG, Triengen, CH)

Pfanzeder scale
→Flat-pan scale (Fig. 120) patented in 1864 by Georg Pfanzeder (*1833, †1910) in which the two →load receptors are each supported by an auxiliary lever in addition to the main lever (Fig. 120b and 120c). Each →platform rests on multiple points, thereby allowing the torque caused by →eccentric loading to be better compensated. This makes

the Pfanzeder scale more robust and less susceptible
to oscillations than, for example, the →Roberval scale.
→Béranger scale

Fig. 120a
Pfanzeder flat-pan scale

a)

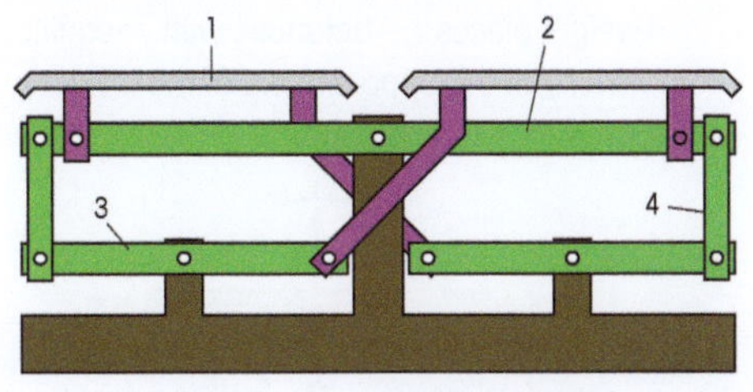

Fig. 120b
Diagram of a Pfanzeder table scale

1: weighing pan
2: main lever
3: auxiliary lever
4: force link

b)

Fig. 120c
Lever system of a Pfanzeder flat-pan
scale

(Images 120a and c by courtesy of
Pfunds Museum Kleinsassen/Rhön,
Hofbieber-Kleinsassen, DE)

c)

pharmacopeia
Officially valid standards and regulations for all prescription
and over-the-counter drugs, food additives, and other human healthcare products, in some cases also animal care
products, including control procedures and applications.
→European Pharmacopeia, →United States Pharmacopeia

pharmacopoeia
Synonym for →'pharmacopeia'.

physical weighing principle
Physical effect on which the →measurement principle of a
→weighing instrument is based. The following effects are
suitable for →weighing and have become technically significant:

A Principles based on the →weight force of the object being weighed (→gravity-dependent weighing instrument)

$$F_G = m \cdot g$$

1. mass comparison

With this measurement principle, the weight force of the load m is compensated by the weight force of reference weights m_r where

$$F_G = m_r \cdot g$$

1.1 Mass comparison with two-arm lever and two weighing pans

Via a lever, the weight force of the load is compensated by the weight force of separate or built-in →weight pieces (→balances with →equilibrium position) at the opposite end of the lever.

1.1.1 Equal arm lever balance (Fig. 121)

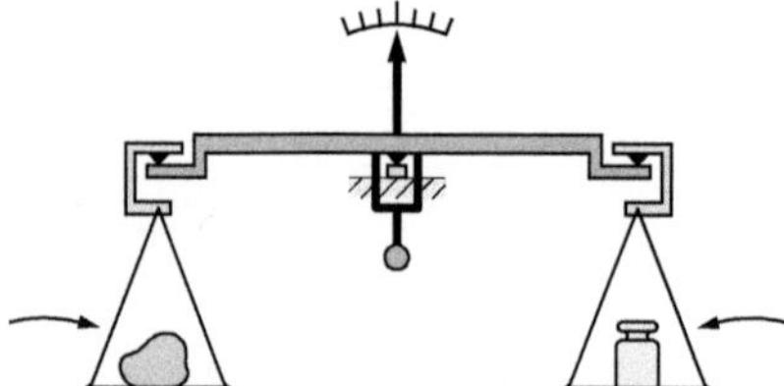

Fig. 121
Principle of the compensating mass comparison (equal-arm lever balance)

1.1.2 →Sliding weight balance (variable →lever arm): Sliding a weight changes the →effective lever arm (Fig. 122)

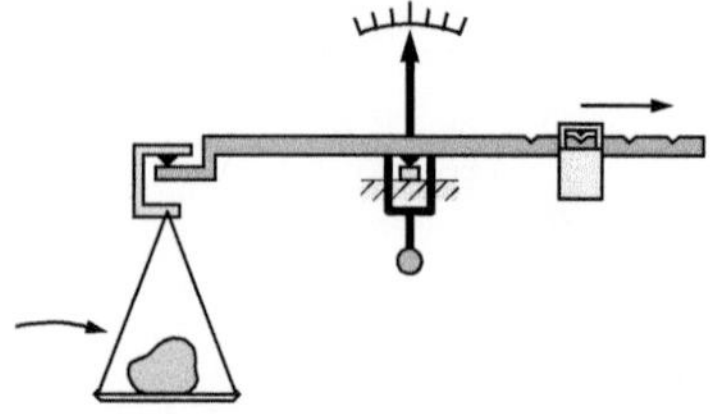

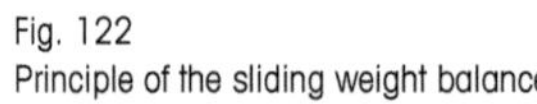

Fig. 122
Principle of the sliding weight balance

1.1.3 →Deflection balance: The effective lever length (→effective lever arm) changes through tilting (Fig. 123).

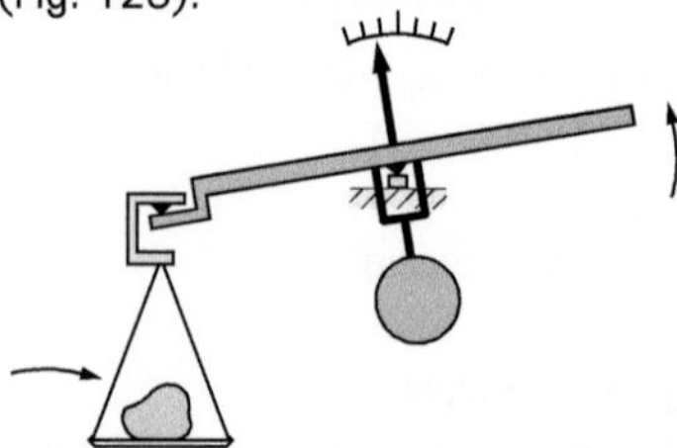

Fig. 123
Principle of the deflection balance

1.2 Substitution principle (one →weighing pan)

The weight force of the load is substituted by removing →weight pieces that are built-in on the load side

(Fig. 124). At the opposite end from the load pan there is an invariable counterweight.

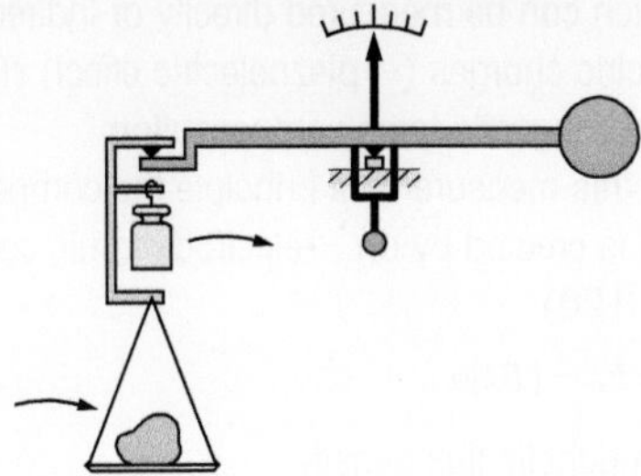

Fig. 124
Principle of the substituting mass comparison (substitution balance)

2. Force comparison
    With this measurement principle, the weight force of the load is not compensated by weight force but by other forces, such as deformation forces or electromagnetic forces

    $$F_G = F_C$$

2.1 Elastic deformation
    With this category, the weight force is guided over a →spring element, which is thereby deformed (Fig. 125). The deformation is a measure of the load. The subsidiary principles differ in how the deformation is measured.

    $$F_C = c \cdot \Delta s$$

    c →spring constant of the deformation body
    Δs elastic deformation

2.1.1 Spring scale: Deformation is measured as displacement (pointer on scale; →projected scale).

2.1.2 Strain gage principle: The deformation (extension and/or compression) causes a measurable change in the electrical resistance of the mounted →strain gage.

2.1.3 Capacitive →converter: The displacement caused by the deformation is measured capacitively (usually by measurement of a differential capacity).

2.1.4 Inductive converter: The displacement caused by the deformation is measured inductively (usually by measurement of a differential inductor or by means of a differential transformer).

2.1.5 Magnetoelastic converter: The deformation (extension and/or compression) causes a change in the magnetic permeability, which causes a measurable change in the inductivity →magnetoelastic effect.

2.1.6 Optical coupler: The deformation is measured optically.

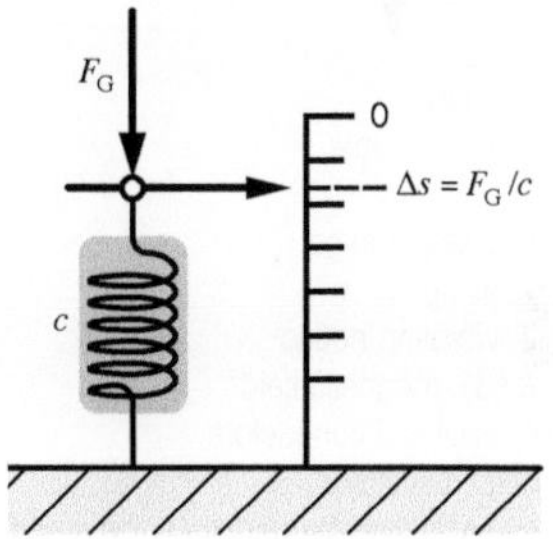

Fig. 125
Compensation with spring body

F_G: weight force
c: spring constant
Δs: elastic deformation

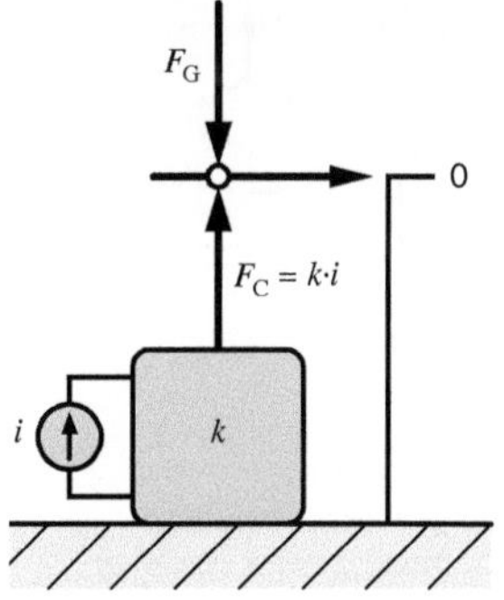

Fig. 126
Compensation with electrodynamic converter

F_G: weight force
F_C: compensation force
i: electric (compensation) current
k: electrodynamic conversion factor

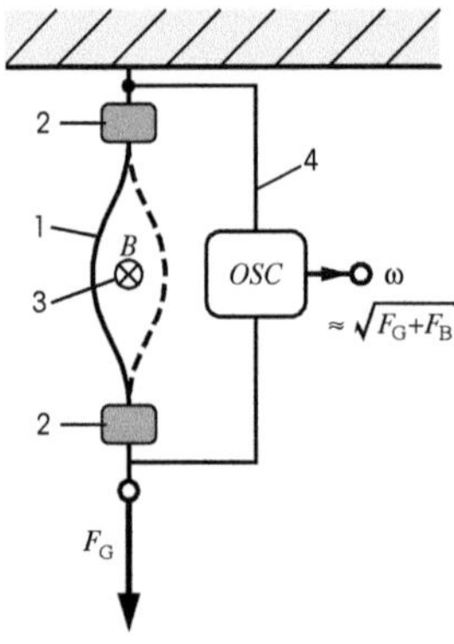

Fig. 127
String principle

F_G: weight force
1: string
2: vibration nodes
B (3): magnetic field
4: electrical conductors
OSC: oscillator
ω: oscillation frequency (of the string)
F_B: buckling force of the string

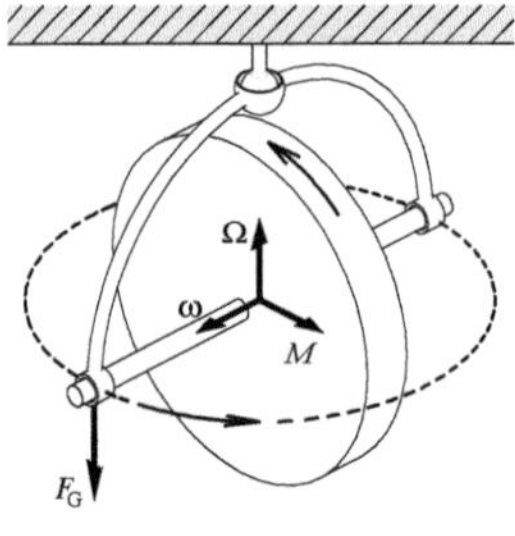

Fig. 128
Gyro principle

F_G: weight force
ω: gyro rotation frequency
M: resulting torque, caused by
loading with the weight force
Ω: gyro precession frequency

2.1.7 Piezoelectric scale: Deformation of a piezoelectric crystal causes →electric charges to be induced, which can be measured directly or indirectly as electric charges (→piezoelectric effect) (Fig. 131).

2.2 Electromagnetic force compensation
With this measurement principle the compensation force is created by an →electrodynamic converter (Fig. 126).

$$F_G = F_C = (Bl) \cdot i$$

B magnetic flux density
l length of the current-bearing conductor in the magnetic field
i electric (compensation) current.

The compensation current is measured. →electromagnetic force compensation

2.3 Vibrating string principle
With this measurement principle, the frequency of vibration of a string as a function of its tension is measured (Fig. 127).

$$f = f_0 \sqrt{\frac{F_G}{F_B} + 1}$$

f_0 resonant frequency of the unloaded string
F_G weight force
F_B buckling force of the string.

Although the compensation force is caused by elastic deformation of the string, it is not measured as such.

2.4 Gyroscope principle
With this →measurement principle, the precession frequency of a gyroscope is measured (Fig. 128); the compensation force is created by the precession torque of the gyroscope (which is not measured).

B Principles that are not based on the weight force of the weighed object

3.   Further measurement principles

3.1 Attenuation of radioactive radiation
With this measurement principle the attenuation of radioactive radiation in a body (→absorption) is measured. The principle is used, for example, for →belt weighers.

3.2 Oscillators
With this →measurement principle, the weighed object is connected to a device that is capable of oscillation. Its resonant frequency depends on the inertia of the sample mass (only suitable for compact samples).

pictogram

A readily understandable stylized pictorial representation for information exchange. In weighing technology, pictograms are standardized in DIN 8125 (Fig. 129).

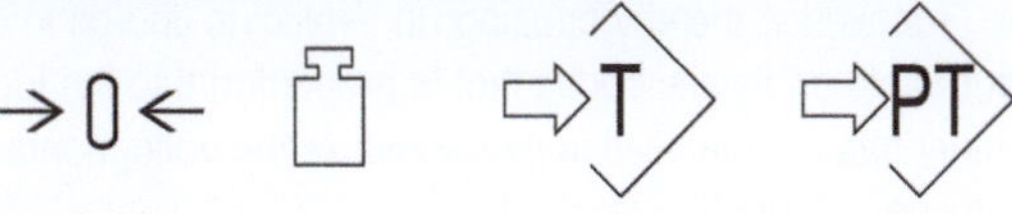

Fig. 129
Examples of pictograms
(from [DIN 8125-1])
Meaning, from l. to r.: zero setting; weight value; set tare; set preset tare value

piece counting

→piece counting system

piece counting system

Measurement system with one or more →counting scales on which, to determine the number of items with the same mass, the total weight of the items is divided by their unit weight. On →mechanical weighing instruments, a piece-counting system consists of several counting pans on one or more →levers with various constant →ratios of mechanical advantage, or of a single counting pan that is mounted movably on a lever which is provided with a counting scale. In the case of →electromechanical weighing instruments, the number of pieces is determined by a computer from the mass of the load and a reference mass (Fig. 130).

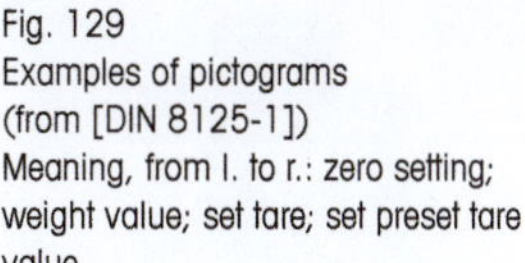

piece-counting device

→piece counting system

Fig. 130
Piece-counting device

piezoelectric effect

Physical effect in which a mechanical force that is exerted on a piezoelectric crystal induces a proportional →electric charge on the opposite faces of the crystal (Fig. 131a). The charge can be measured either directly or via the electric potential that it causes. Typical materials used are, for example, quartz or other piezoelectric monocrystals (Fig. 131b).
→physical weighing principle, →piezoelectric scale

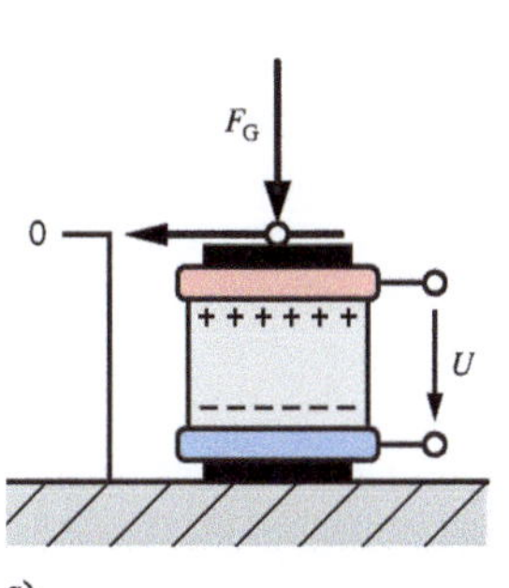

Fig. 131
a) Piezoelectric effect;
b) Piezoelectric monocrystals

(Image 129b by courtesy of Kistler Instruments AG, Winterthur, CH)

Fig. 132
Pin strain-gage load cell for high
loads with a weighing capacity of up
to 500 t

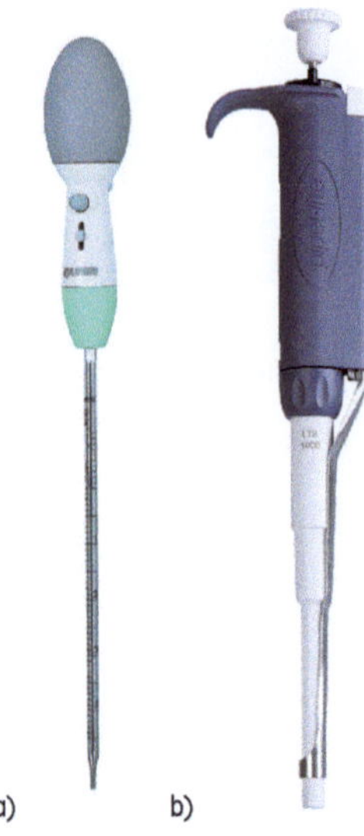

a) b)

Fig. 133
Pipettes
a) measurement pipette made of
glass with rubber suction bulb;
b) manual pipette with replaceable tip

piezoelectric scale

Weighing instrument whose →physical weighing principle
is based on the →piezoelectric effect. The →weight force
acts on the piezo crystal (Fig. 131b), which functions as
a →converter, thereby causing an →electric charge to accumulate on its electrodes that is proportional to the load.
Either this charge itself is measured, or the voltage that it
causes between the electrodes.

pin load cell

A.k.a. compression column load cell or rocker pin. →Strain-
gage load cell with a cylindrical steel rod as →spring element that is compressed by the load (Fig. 132). Since the
spring element is very stiff, these cells are used for weighing
instruments with a high →weighing capacity, e.g. →road
vehicle scales. →rail scale

pipette

A tube into which quantities of liquid can be taken up, or
from which quantities of liquid can be ejected, by the application or release of suction. A simple pipette consists of
a glass tube that has a volume →scale printed on it, has a
point at its lower end, and at its upper end is either open,
or closed with a rubber squeezer (Fig. 133a) (→measure-
ment pipette). Manual pipettes (also called liquid handling
instruments), have a piston, whose suction volume can be
adjusted (Fig. 133b), that draws the liquid in either directly
(so-called direct displacement), or indirectly via an air volume (so-called air displacement), and subsequently ejects
it. Pipettes are suitable for dispensing very small volumes
of liquid (microliters). To prevent contamination, pipettes
are usually fitted with replaceable tips. ISO Standard 8655
includes requirements for the construction and testing of
pipettes using →gravimetric methods to determine their
measurement uncertainty ([ISO 8655-6]).

pivot joint

Movable coupling (→joint) that guides mechanical parts
relative to each other but allows a rotating movement. Pivot
joints may comprise, for instance, pins, ball bearings, cam
and steel tape, →knife-edge with →pan, or elastic elements
such as →flexible joints or →flexible couplings.

place of installation

1. Place where a weighing instrument is set up and used.
   (Compare: →place of use)
2. →Support

place of use

Location or zone of use of a weighing instrument in which the value of →gravity is assumed to be constant. The place of use may be a city, an administrative area, or larger region, provided that gravity can be assumed to be sufficiently constant throughout the location. →place of installation, →zone of use

place of verification

Official testing laboratory of the →Weights and Measures authority, →zone of use, →place of use, or →place of installation of the measuring instrument. Verification at the place of installation is required when

1. on-site verification is expressly stipulated in the legal metrology requirements or in the approval (e.g. for high-resolution weighing instruments);
2. there is a danger that the measuring instrument could be damaged in transit;
3. transport of the measuring instrument is not possible;
4. the measuring instrument is connected to other instruments or devices on site so that it can be assessed only together with such devices.

platform

A device that serves to accommodate the load (→load receptor) on larger scales, usually executed low and flat. Compare: →pan, →weighing pan

platform scale

→Bridge scale for industrial use with flat →load receptor (→floor scale, →low-profile scale) and usually ground-level bridge surface (→rail scale, →road vehicle scale).

platter

→pan

PLU

Abbreviation for 'Price Look Up'. Data memory built into a →counter scale for →base prices which, for instance, can be called up for a particular article by pressing a key.

plumb line

Device used to indicate the horizontal installation or reference position of a weighing instrument. The device consists of a downwardly pointed conical pendulum that hangs over a fixed marking. →inclination

plummet
→sinker

plunger
A →displacement body, usually made of metal, and spherical in shape, for measuring the density of liquids, that is held on a rod and immersed in a liquid to determine its density (→gamma sphere) (Fig. 74). Since the volume of the plunger is known, the density of the liquid can be determined directly from the →buoyancy (→density determination). →sinker

plus/minus balance
→over/under scale

point of sale, public
→public point of sale

pointer
The movable reading element of a →mechanical weighing instrument with a fixed →scale. →analog readout

poise beam
→rider system

poise weight
Common term in the United States of America for →'rider'.

position sensor
Electromechanical converter that converts the position of an object into an electric signal (Fig. 134). Position sensors make use of, for example, a differential condenser, a differential transformer, or an electro-optical converter (→position vane).

Fig. 134
Electro-optical position sensor of an electrodynamically compensating load cell (→EMFC load cell).
Visible at the left of the picture is the light source, in the middle the movable vane, and at the right the light receiver that receives the light modulated by the vane and converts it into an electric position signal.

position vane
Moving part of an electro-optical →position sensor. The position vane usually has a slit through which the light emitted by the sensor passes. The amount of light that passes through the vane is determined by the position of the vane, and modulates the electrical output signal. In the case of electromagnetically compensating weighing cells (→EMFC load cell), the position vane is usually fastened to the →lever.

postal rate indicating machine
Instrument combination comprising a scale with EDP and a printing device, if necessary also with supply and removal devices with which postal rates for parcels, etc. can be cal-

culated, printed on a prepaid stamp (label), and stored and evaluated for accounting and other statistical purposes.

postal scale
→Scale with weight ranges used to determine the weight of a letter or parcel and the price for its mailing.

power failure protection
A device or measure that signals a temporary or permanent power failure, or prevents the →indication, →printout or release of an incorrect measurement value, for example a built-in storage battery that serves as an emergency power supply.

PPD
Abbreviation for →'Prepackaged Products Directive'.

precision
1. Qualitative term describing the scatter of measurements.
2. The closeness of agreement between independent →measurement values obtained under stipulated conditions ([ISO 5725] 3.12) (Fig. 1).

Precision depends only on the distribution of the →random error, not on the true value of the measurand (→trueness). Example: The ability of a measuring instrument to provide measurement values with little scatter. →accuracy (compare: →trueness)
Note: The precision can only be determined when multiple measurement values are available.

precision balance
A →weighing instrument of high accuracy (Fig. 135).

precision weight
1. →Weight piece of the medium →error limit class. →weight class
2. Term defined in the German →Verification Ordinance for →weight pieces of class OIML F1 (→OIML weight classes) ([VO] Appendix 8, Section 2, 2.1).

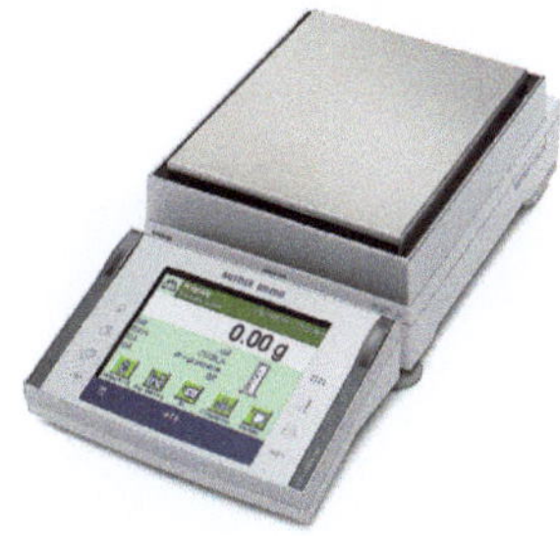
Fig. 135
Precision balance with a weighing capacity of 4 kg and a readability of 0.01 g

prepackage
Within the meaning of the →Prepackaged Products Directive, prepackages are products in any type of package that have been packaged, sealed, and placed on sale in the absence of the purchaser. It is impossible to change the quantity of the product contained therein, which has a consistent, predefined value (nominal fill quantity) without opening or otherwise noticeably changing the package. Prepackages

must comply with certain requirements that are stipulated in the →Prepackaged Products Directive or →Prepackaged Products Decree. →e-mark

prepackage process control
The duty of the manufacturer, as stipulated in the →Prepackaged Products Directive, to verify by means of random sampling (→statistical quality control) the →fill quantity of the →prepackages that it manufactures and to record the results (→obligation to record). →Prepackaged Products Decree, →e-mark

Prepackaged Products Decree (PPD)
A legal ordinance valid in Germany that is based on the →Weights and Measures Act and converts the European →Prepackaged Products Directive into national law. It contains rules and regulations for products in →prepackages, especially requirements for →nominal fill quantities and container volumes, minus deviations, fill quantity and base price indication, as well as verification of the filled packages (→prepackage process control).

Prepackaged Products Directive
The European Prepackaged Products Directive →76/211/EEC applies to →prepackages. It is binding for prepackages with a →nominal fill quantity between 5 g and 10 kg or 5 ml and 10 l. For example, according to the Prepackaged Products Directive, the mean value of the actual fill quantity of all prepackages of a batch must be not less than the stated nominal fill quantity, and certain negative deviations must not be fallen below. The manufacturer has a duty to measure the actual fill quantity with a suitable measuring instrument and to record the results. The measuring instruments that are used are subject to the applicable measurement laws (e.g. →Measuring Instruments Directive). The requirements stipulated in the Prepackaged Products Directive are implemented as national law in the EEA and Switzerland (→Prepackaged Products Decree). For compliance with the Prepackaged Products Directive, use of the →e-mark is required (Fig. 52). The regulations for monitoring by the national authorities vary from country to country. →76/211/EEC

prescription balance
A →weighing instrument of high accuracy or →weighing instrument of special accuracy that is particularly suitable for use in pharmacies on account of its maximum capacity, accuracy, and ease of operation (Fig. 136).

Fig. 136
Prescription balance with mortar

preset tare device
Device with which a prescribed tare value can be subtracted
from the →gross weight or →net weight and the calculated
result indicated ([OIML R 76-1] T.2.7.5). The →weighing
range for →net loads is reduced by the tare value.

pressure
Physical quantity that expresses the force acting per unit of
area. The SI unit of pressure is the pascal: $1 \text{ Pa} = 1 \text{ N/m}^2$;
the former unit bar ($1 \text{ mbar} = 1 \text{ hPa}$) is still also in use.

preweighing
The →weighing of a →weighed object for the purpose of
approximately determining its mass, the definitive →weigh-
ing result being determined with a more accurate weighing
instrument.

price indicator
Additional indicating device of self-equilibrating, price-computing weighing instruments. The device serves to indicate
the →purchase price of the weighed quantity of merchandise
and is only permitted to be used together with an indicating
device that visibly displays the applicable weight price and
the base price at one and the same time.

price marker
Price marker scales are used for marking the price on prepackaged goods. →price marker

price marker scale
→Scale used in the production of →prepackages of unequal
→fill quantity comprising a →self-indicating instrument,
computer, and printer device. Weight, base price, and purchase price of the prepackage are printed simultaneously
on a usually self-adhesive label. →Prepackaged Products
Directive

price-computing weighing instrument
Instrument that calculates the price to pay on the basis of
the indicated weight value and the unit price ([OIML R 76-1]
T.1.2.8).

primary display
Primary displays are displays for weighing results, the
correct zero position (zero indicator device), and for operation of the tare device. In public point of sale they also
indicate the base price and purchase price, purchase price

for unweighed articles, and possibly also number, price per item, and total price ([OIML R 76-1], T.2.2.6). →auxiliary indicator

print lock
Device attached to weighing instruments to prevent a weighing result from being printed when influences are at work that would falsify the weighing result. →stand-still lock, →stand-still detector

printed record
Printouts from verified printer devices must indicate whether the printed weighing values are weighing results that were produced by a verified weighing instrument or whether they are calculated weighing values. Values other than weighing values must be indicated by the respective unit or its symbol or other special character.

printer
→printer device

printer device
That part of the printing apparatus that transfers characters onto paper or other media.

printing
Method of reproduction for transferring information such as text and images onto a carrier material such as paper. →printout, →printer device, →alibi printer

printing device
Device that prints the weighing result onto, for example, paper, cards, lists, or reeled paper tape. Certified printers for weighing instruments consist of →stand-still locks, possibly also measurement value converters and control components, connecting cables, and the actual →printer device.

printout
Weighing result or other data printed out in the form of a record by a →printer device (Fig. 137).

proFACT
Abbreviation for 'Professional FACT' (vendor-specific name), a designation for the →automatic adjustment of the sensitivity. The points in time at which an adjustment should be performed can be specified by day of the week and time of day. →Autocal, (→FACT)

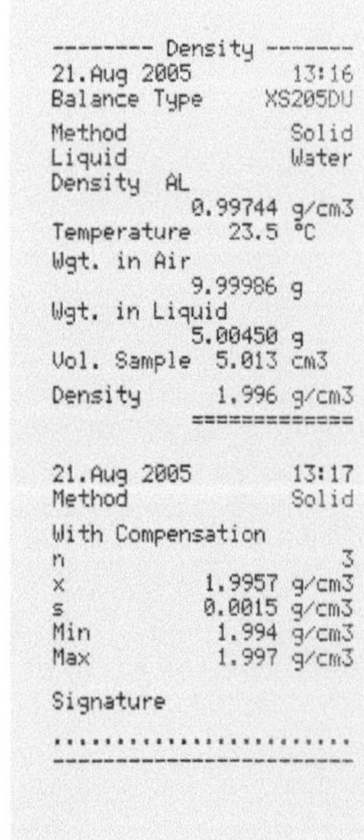

Fig. 137
Printout

program return after power failure
Specified restart of the program, e.g. with the memory content (variable) before the disturbance, restart of the program at the former position.

programmable or loadable software
→Legally relevant software in weighing instruments, personal computers (PC), weighing instruments with PC assemblies, and other devices or →modules such as →control units, →terminals, →data memories, →auxiliary devices, etc. that can be loaded after verification. The software must comply with the rules for legally relevant software and software separation (verified / not verified). →software securing, →software identification

projected scale
A →scale image or section of a scale image on a →mechanical weighing instrument that is either projected onto a screen or shown on a fluorescent screen equipped with an equilibrium indicator.

proportional weighing method
Simple weighing in which the weighed object is placed on the →load receptor (→load pan) and the →mass (→weight) read off.

protected interface
→Interface (→hardware and →software) across which only such data can be entered into the data processing device of a →weighing instrument, a →module, or an →electronic assembly that
a) do not produce readouts that are not clearly defined and could be confused with →weighing results;
b) cannot falsify →weighing results that are displayed, processed or stored, or →primary displays;
c) cannot →adjust the weighing instrument or change adjustment data. This does not apply to adjustment with built-in devices or with external adjustment weights on weighing instruments of →accuracy class ①.

protection type
→degrees of protection provided by enclosures

prototype
1. Original, sample, first printing.
2. Mass unit: In the hierarchy of mass normals, the highest mass normal. →International Prototype of the Kilogram

a)

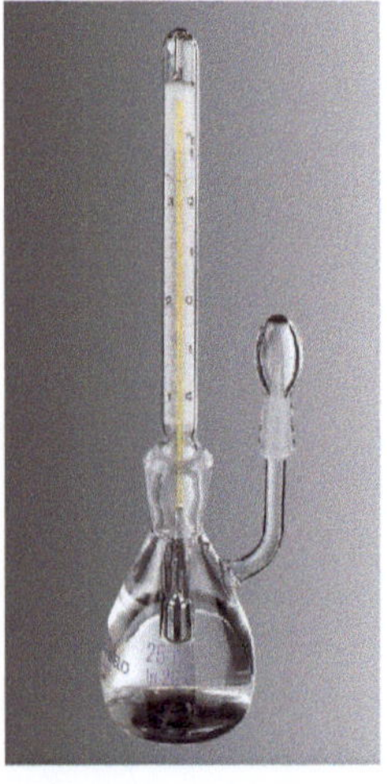

b)

Fig. 138
Pycnometer
a) simple type;
b) with integrated thermometer

(Images by courtesy of
Paul Marienfeld GmbH & Co. KG,
Lauda-Königshofen, DE)

Fig. 138c
Pycnometer for determining the
density of OIML E & F weights
up to 20 kg

PTB
Abbreviation for 'Physikalisch-Technische Bundesanstalt'
(Federal Institute of Physics and Metrology), the German
→national metrology institute with headquarters in Braunschweig and Berlin (www.ptb.de).

public point of sale
Point of sale accessible to everyone where the sale or purchase of goods is carried out, or measurable services provided against payment. Typical public points of sale are, for
instance, shops, market stalls, kiosks. Additional regulations
are in force for scales in public points of sale stipulating,
for example, displays for purchaser and seller, indication of
weight, price, and →base price.

public scale
A →weighing instrument on which public weighments are
performed, i.e. products are weighed for anyone. The results
of the weighments are certified by a publicly appointed and
notarized →weigher (weighmaster) or a trading standards
officer.

purchase price
The price of the weighed object as the product of its determined mass and the →base price.

pycnometer [15]
A glass vessel of exactly defined and constant volume used
to determine the →density of liquids, solid bodies (→weight
pieces), or insoluble bodies (e.g. powders) (Fig. 138).

c)

---

[15] pyknos (Greek): thick

quality
The entirety of properties and characteristics of a product
(or of a service) that relate to its suitability for fulfilling
specified requirements. Quality characteristics of measuring
instruments include their measurement trueness (i.e. their
adherence to specified →error limits), their →measurement
uncertainty, and their →reliability.

quality assurance
All planned and systematic activities that are needed to create appropriate confidence that a product (or a service) will
meet certain quality requirements ([ISO 9000] 3.2.11).
These quality requirements can be defined by the manufacturer of the actual product or can be specified in standards
or other documents pertaining to standards.

quality control
The operational techniques and activities that are used to
fulfill specified quality requirements and eliminate non-
conforming results ([ISO 9000] 3.2.10).

quantity counting device
A counting device that is most frequently used on →auto-
matic weighing instruments for →weighing and apportioning (→apportion) that indicates the number or quantity (in
→units of mass) of fillings.

rail scale

→automatic rail scale

rail wagon scale

→automatic rail scale

random deviation

→random error

random error

1. Component of measurement error that in replicate measurements varies in an unpredictable manner. ([VIM:2008] 2.19)
2. Deviation of the uncorrected measurement result from the expected value. ([DIN 1319-1] 3.5.1)
3. A measure of the scatter of measurements. Example: Deviations that are caused by changes in the weighing instrument that cannot be identified or influenced, in interference quantities, in the weighed object, and in the observer (→weigher). These cause a scatter in the weighing result, which can be quantified by the application of statistical methods. →precision

range displacement

Device on a weighing instrument to displace the measurement range without changing the sensitivity. The range is displaced by switching a built-in weight in or out or, in the case of electromechanical weighing instruments, by an electric signal.

range switching

Can be done by

1. switching a built-in weight in or out;
2. changing the sensitivity of the load cell of an electromechanical weighing instrument; or
3. changing the evaluation device.

→multi-range weighing instrument

rapid drying procedure

With the rapid drying procedure, →dryers measure the →moisture content of a sample quickly and easily. In contrast to the →drying oven method, in which the drying result is determined in a complex sequence of work steps lasting hours or days, in the dryer, the heating unit and weighing unit are combined in the same instrument, which produces quick results.

rate indicating scale
A →scale used to determine fees such as transportation fees. In addition to the weight indicator, a rate indicating scale also has a fee indicator.

ratio of mechanical advantage
Ratio of the effective lengths (→effective lever arm) of the two arms of a →lever. In the case of the connected levers of a →lever chain, the term is applied to the product of the various ratios of mechanical advantage.

readability
1. →Specification: The smallest difference in mass that can be read on a weighing instrument. For instruments with a digital display, the readability is equal to the division value →actual scale interval (→digital interval) of the display. For weighing instruments with a scale indicator, the readability is the smallest fraction of a division that can still be estimated with reasonable reliability (in the case of an analog indicator, for example, 0.2 scale intervals) at the usual reading distance, or which can be determined with the assistance of an auxiliary device (→fine adjuster). Expressed in units of mass, e.g. [g].
2. The minimum height of the numbers on the display or weighing-out device to ensure unmistakable readability.

readiness
A term used in Weights and Measures regulation in Germany. Readiness exists, and therefore the obligation to certify, if applicable, when the weighing instrument can be used without special preparations.

readout error
Obsolete term for →display error.

readout stabilization
→Electronic device that keeps the indicated value stable even though the internal measurement values are impaired by environmental influences such as vibrations. →filter

receiving scale
A →scale used to determine the mass of incoming goods.

reference current
Constant reference current for analog-digital conversion (→analog-digital converter).

reference density
Conventional value for density that is specified for definition
of the →conventional mass and is
1. for the reference weight: 8000 kg/m³;
2. for air: 1.2 kg/m³.
[OIML D 28]

reference mass
→Mass standard, often in the form of a →weight piece, that
is used as a reference quantity for the →calibration or →ad-
justment of weighing instruments or other mass standards.
→standard, →standard weight

reference method
1. →Method that deviates only slightly from the required
   accuracy. The accuracy of a reference method must be
   demonstrated through direct comparison with the definitive method. [IUPAC]
2. Measurement method for determining the →moisture
   content that allows →traceability to (legal) standards.
   Depending on the reference method, different components
   of the moisture content (free, bound, crystalline water)
   can be included in the measurement result.

reference position
→reference position of the weighing instrument

reference position of the weighing instrument
Geometrical position of the weighing instrument which is
aligned with its →axis of action, and at which its operation
is adjusted ([OIML R 76-1] T.6.4). →inclination, →level
indicator

reference voltage
Constant reference voltage for analog-digital conversion
(→analog-digital converter).

reference weight
1. Synonym for →'reference standard'. →standard,
   →weight piece
2. →Reference mass for →adjustment or →calibration of
   the sensitivity or other characteristics (e.g. →linearity) of
   a →weighing instrument. The reference mass may take
   the form of an external →weight piece or a weight that
   is built into the weighing instrument (Fig. 139) (→self-
   adjustment). Whereas an external reference weight is
   traceable, a built-in weight is not; the effect of a built-in

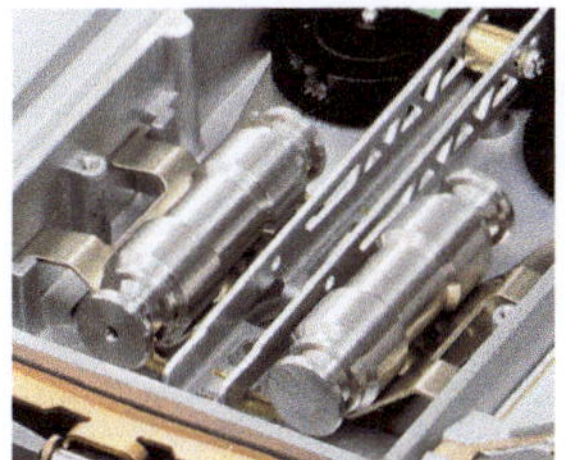

Fig. 139
Reference weights built into the
weighing instrument for adjusting the
sensitivity

weight can, however, be traced by comparing its effect with an external reference mass (→traceability).

3. A representative reference mass of a single item for →piece counting. For small and light items, the reference weight can be determined by weighing a multiple (10 to 100).

relative resolution

Ratio between the →scale interval d and the →maximum capacity Max (= reciprocal of →number of scale intervals n_d)

$$res = \frac{d}{Max} = \frac{1}{n_d}$$

reliability

The ability of a measuring instrument or its individual components to operate with certainty according to the requirements for a defined period of time. Reliability is a quality characteristic and is achieved by means of proper design, →quality assurance and maintenance.

repairer identification mark

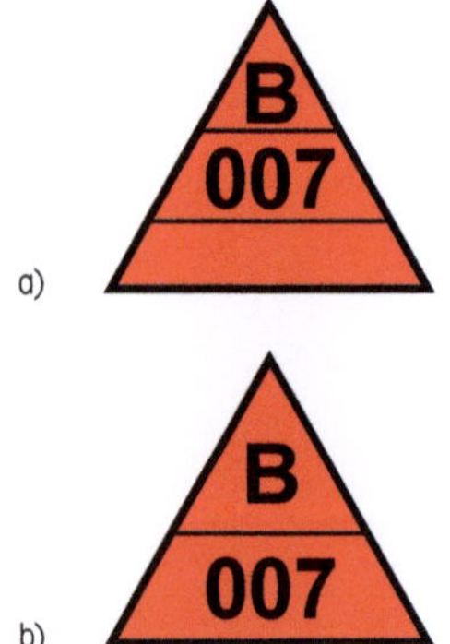

a)

b)

Fig. 140
a) Repairer identification mark;
b) Repairer identification stamp

The repairer identification mark valid in Germany is a triangular sticker (Fig. 140). The upper field contains the identifying letter of the →Weights and Measures authority (→ordinal number), that issued the mark, while the middle field contains the number assigned to the corrective maintenance technician. The bottom field is reserved for the date of the corrective maintenance. The background color of the sticker is signal red, the lettering is black. In other countries of the EEA, and in Switzerland, there are also repairer identification marks, whose formats and inscriptions differ from country to country.

repairer identification stamp

→repairer identification mark, Fig. 140b

repeatability

1. The degree of agreement between the results of successive measurements of one and the same measurement variable, conducted under identical measurement conditions.

   Note: The conditions are referred to as repeatability conditions ([VIM:2008] 2.20).

   Repeatability conditions are
   – the same measurement procedure
   – the same observer
   – the same measuring instrument
   – the same place
   – a short time between measurements.

The repeatability can be expressed quantitatively as the scatter values of the results.
(Compare: →reproducibility)
2. Ability of a weighing instrument to provide results that agree one with the other when the same load is deposited several times and in a practically identical way on the load receptor under reasonably constant test conditions ([OIML R 76-1] T.4.3).
The measurement series must be performed without interruption by the same operator using the same →weighing method in the same position on the →load receptor at the same →place of installation under constant environmental conditions (→environmental influence).
The →standard deviation of the measurement series is a suitable measure to determine the value of the repeatability. Alternatively, it can be expressed as the difference between the largest and smallest measurement value of the measurement series ([OIML R 76-1] 3.6.1).
2.1 →Specification: As in 2. The repeatability is usually expressed as the standard deviation and, unless stated otherwise, relates to
A) the →net value of the →weighed-in quantity or →weighed-out quantity (not the tare or gross value),
B) an individual weighing (not to the mean value of a measurement series).
Expressed in units of mass, e.g. [g].

reproducibility

Extent of the agreement between the →measurement values of the same →measurand when the individual measurements are performed under different conditions ([VIM:2008] 2.24) with regard to, for example,
– the measurement method
– the observer
– the measuring equipment
– the measurement site
– the application conditions
– the point in time.
Note:
1. A statement of the reproducibility must be accompanied by information about the different conditions.
2. The →standard deviation of the measurement values is a suitable measure for the value of the reproducibility.
(Compare: →repeatability)
3. Ability of a weighing instrument to display identical measurement values for repeated weighings of the same

object under different conditions. The conditions that changed must be stated. These could be, for instance, the operator, →weighing method, position on the →load receptor, →place of installation, →environmental influence, or interrupted operation.

requirements for measuring instruments

In legal metrology, measuring instruments must satisfy certain legal and technical requirements, e.g. →type approval, →inscriptions, →error limits, →Verification Ordinance. Additional requirements regarding →electrical safety and →electromagnetic compatibility must also be fulfilled. →Measuring Instruments Directive, →Directive on Non-Automatic Weighing Instruments, →EMC Directive

resolution

1. Smallest difference between displayed indications that can be meaningfully distinguished. ([VIM:2008] 4.15)
2. Quantitative specification of the capability of an instrument or a displaying device to distinguish unequivocally between →measurement values that lie close to each other. →readability, →actual scale interval, →relative resolution
3. Non-technical expression for →number of scale intervals.

response threshold

→discrimination

rest position

Synonym for →equilibrium position.

rider

Irremovable →weight piece (a.k.a. poise weight or counterpoise weight) that can be moved on a poise beam ([OIML R 76-1] T.2.5.1). The position of the rider in relation to a scale indicates the →weighing value. →sliding weight balance

rider system

→Weighing-out device consisting of one or more poise beams that maintain the →equilibrium of a →sliding weight balance. Characteristic is the sliding →rider on the poise beam that is graduated with notches or lines (Fig. 141). Moving the rider changes the →effective lever arm. Thus each position of the rider on the poise beam corresponds to a specific →weighing value.

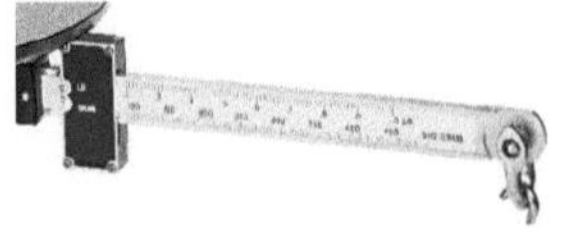

Fig. 141
Rider system with rider (poise weight) and poise beam

ring weight

A ring-shaped →weight piece used primarily in →dial weight balances. If the lifting mechanism is appropriately designed, ring weights reduce disturbing oscillations of the →hanger and of the →load receptor. Different sizes of ring weights can be arranged so that their respective centers of gravity coincide, which is advantageous in avoiding →eccentric loads.

road vehicle scale

→Vehicle scale for road vehicles, a.k.a. truck scale, executed as a →bridge (Fig. 142) or as a →dynamic axle-load scale (Fig. 51). →rail scale

Fig. 142
Road vehicle scale

Roberval scale

Flat-pan →scale with →parallel guide that was invented in 1669 by Gilles Person(n)e de Roberval (*1602–†1675) (Fig. 143a). This construction was the first in which the →load receptors of a scale were arranged in →top-loading manner instead of in the formerly usual hanging manner (Fig. 143b). The principle is still used today for guiding the load receptor of low-loading weighing instruments. →parallelogram

Fig. 143a
Roberval scale

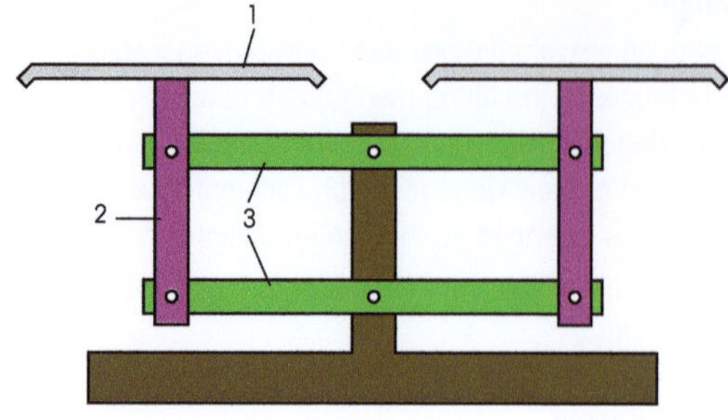

Fig. 143b
Diagram of a Roberval scale

1: load receptor
2: hanger
3: lever

Fig. 144
Roman dial scale from Augusta
Raurica (Augst), CH
(Image by courtesy of the Augst
Roman Museum, Augst, CH)

rocker pin
→pin load cell

Roman beam scale
→Roman dial scale

Roman dial scale
The oldest version of the simple →sliding weight balance
with one →rider (Fig. 144).

rope-tension scale
A →scale used →to weigh a load hanging from a rope by
determining the traction of the rope. The mass of the rope,
which changes with the length of the rope as the load is
raised and lowered, and if applicable also the influence of
the various positions of the crane jib, are automatically compensated. →crane scale

rounding error
1. Deviation that occurs when a value is rounded off before
   it can be indicated by a digital display. If rounding takes
   place upwards when the last digit is 5 or more, and otherwise downwards, the rounding error lies between plus/
   minus half a →scale interval d and is evenly distributed;
   the expected value (→mean value) of the rounding error
   is zero, and its standard deviation is $d / \sqrt{12} \approx 0.3\, d$.
   The standard deviation of the difference between two
   numbers that are rounded to the same number of decimal places is $\sqrt{2}\left(d / \sqrt{12}\right) \approx 0.4\, d$.
2. Error of a →digital display that cannot be measured
   directly but would be visible in an →analog readout
   ([OIML R 76-1] T.5.4.3). →rounding of measurement
   results

rounding of measurement results

1. Measurement results should be rounded to the same number of significant digits as the measurement uncertainty which, when expressed in mg, is rounded to two significant digits ([GUM] 7.2.6).
2. Weighing instruments that have a digital display round the internal measurement value to the →readability (→scale interval) upwards when the last digit is 5 or more, and otherwise downwards. This results in a →rounding error.

salesperson keys
Function and input keypad of a →counter scale that is located on the salesperson side of the scale. (compare: →customer keys)

sample
1. Sample (subset) of a total population that is taken when a characteristic of the total population should be determined but the effort of inspecting all elements of the total population would be excessive. The sample must be chosen in such manner that it is representative of the total population.
2. A term used in →prepackage process control to designate →prepackages that are removed for weight checking.
3. →Weighed-in quantity of a substance.

sample size
Number of units in a →sample.

scale
1. →Weighing instrument, intended predominantly for medium to high capacity →weighments, with moderate to low resolutions, used indoors or outdoors in office and industrial environments, and typically of OIML class (III) or (IIII). →strain gage scale, →vehicle scale, →monorail scale, →tank scale, →air baggage scale, →bathroom scale
2. Sequence of →division marks, dots, or numbers on a dial (Fig. 145).

Fig. 145a
Scale

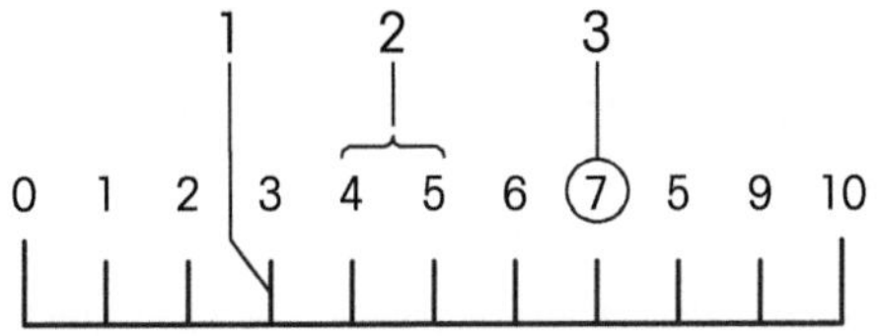

Fig. 145b
Scale

1: scale mark
2: scale division
3: scale value

scale cash register
Cash register with attached →checkout scale, allowing entry of the →base prices (→PLU) and →printout of the weight, base price, and price, at the cash register. →cash register systems

scale division

Smallest increment of a →scale (Fig. 145b, 2). →scale
interval, →readability

scale interval

Generic term for the difference between two consecutive
scale marks (analog indication) or indicated values (digital
indication). →actual scale interval

scale intervals, number of

→number of scale intervals

scale mark

→Division mark or notch of an evenly divided, graduated
→scale (Fig. 145b, 1) ([OIML R 76-1] T.2.4.2).

scale pit

Pit that accommodates a →bridge scale whose bridge surface is usually level with the ground (Fig. 146).

scale spacing

The distance between adjacent →division marks, measured
along the base line of the scale in length units.

scale value

Value of a →division expressed in mass units
(Fig. 145b, 3). →display

Fig. 146
Scale pit

scanning device

A device used to determine the change in the position or
angle of the →weighing-out device of a balance caused by
the load, primarily by using mechanical, electrical, or optical
sensors. →position sensor

seal

Term used in non-technical language for →sealing.

seal, to

1. In the broader context: Application of a verification seal
   (→sealing point, →software securing).
2. In the specific context: The securing of metrologically
   important parts against displacements, alterations, or
   removal by the attachment of metallic objects (seals) or
   adhesive seals. A seal is stamped on these objects by
   means of pliers (lead-sealing pliers).

sealing

Seals the parts contained in a →housing against prohibited interventions, usually by means of a stamp. →to seal

sealing point

Components of verifiable weighing instruments that should not be adjusted or removed by instrument users must have appropriate locations where securing seals relating to verification (e.g. lead seals, verification marks) can be attached. Under certain conditions, sealing by →software (→software securing) is also possible.

securing sticker

Self-adhesive label of paper or plastic used, for example, for the →main verification mark, →data plate, or →sealing.

sedimentation balance

A weighing instrument designed to determine the particle size of sediments. The instrument registers the mass of the particles (sediment) deposited on the weighing pan as a function of time. From a curve obtained in this manner, it is possible to determine the particle distribution.

self-adjustment

→Automatic adjustment of a weighing instrument with a built-in adjustment device that contains a →standard (usually a →reference mass, possibly also a →reference voltage, or similar).

self-equilibrating instrument

→self-indicating instrument

self-indicating instrument

Weighing instrument in which the position of equilibrium is obtained without the intervention of an operator ([OIML R 76-1] T.1.2.3). →semi-self-indicating instrument, →non-self-indicating instrument

self-indication capacity

Weighing capacity within which equilibrium (→settling position) is obtained without the intervention of an operator ([OIML R 76-1] T.3.1.3).

self-service weighing instrument

→Scale in a public point of sale that is intended for use by customers and usually produces a receipt showing weight, price, base price, and type of goods, that can be attached to the weighed goods (Fig. 147) ([OIML R 76-1] T.1.2.10).

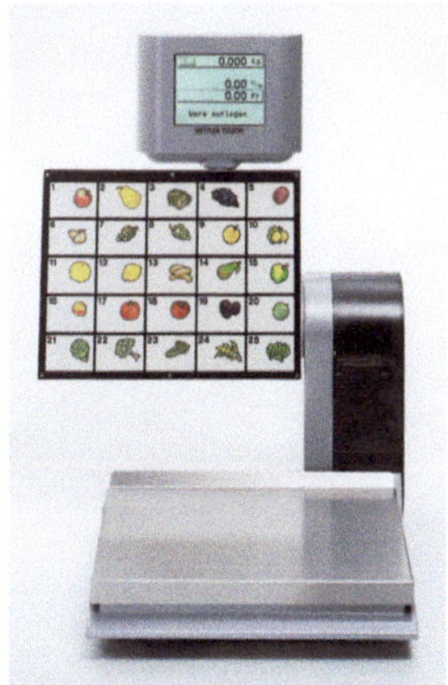

Fig. 147
Self-service weighing instrument

semi-self-equilibrating instrument
→semi-self-indicating instrument

semi-self-indicating instrument
Weighing instrument with a self-indicating weighing range, in which the operator intervenes to alter the limits of this range ([OIML R 76-1] T.1.2.4). →self-indicating instrument, →non-self-indicating instrument

semimicro balance
→Analytical balance with a typical →weighing capacity of between 50 g and 200 g and a →readability of 0.01 mg (Fig. 148). →weighing instrument of special accuracy

Fig. 148
Semimicro balance with weighing capacity of 200 g and readability of 0.01 mg

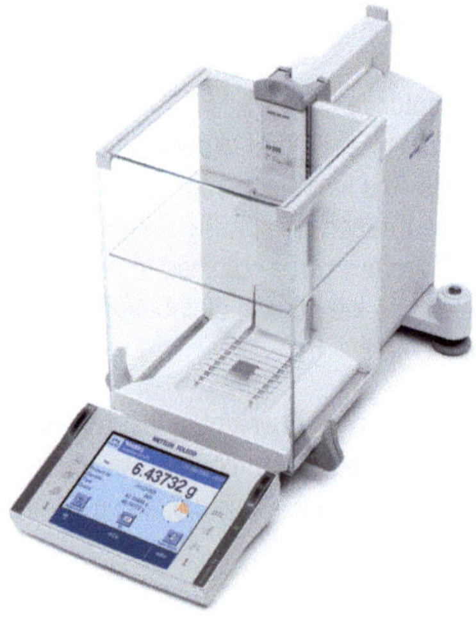

sensitivity
Change in the output variable of a measuring instrument divided by the associated change in the input variable ([VIM:2008] 4.12; [OIML R 76-1] T.4.1). In the case of a weighing instrument, the change in the →weighing value ΔW divided by the change in load Δm that causes it

$$S = \frac{\Delta W}{\Delta m}$$

(differential sensitivity). If the change in the measurement value is expressed in mass units, the sensitivity is a dimensionless quantity whose correct value is 1.
The sensitivity is one of the most important →specifications of a weighing instrument. The specified sensitivity of a weighing instrument usually relates to its global sensitivity (slope) measured over the nominal range

$$S = \frac{\Delta W_{\mathrm{nom}}}{\Delta m_{\mathrm{nom}}} \quad \text{(Fig. 149)}.$$

A deviation of the →characteristic curve from the straight line of the global sensitivity is specified via the linearity deviation (→linearity).

The sensitivity of weighing instruments whose weighing principles (→physical weighing principle) are based on the measurement of the weight force is proportional to →local gravity. Depending on the →number of scale intervals of the weighing instrument at its →place of use or →place of installation, the sensitivity must be adjusted (→sensitivity, →sensitivity adjustment).

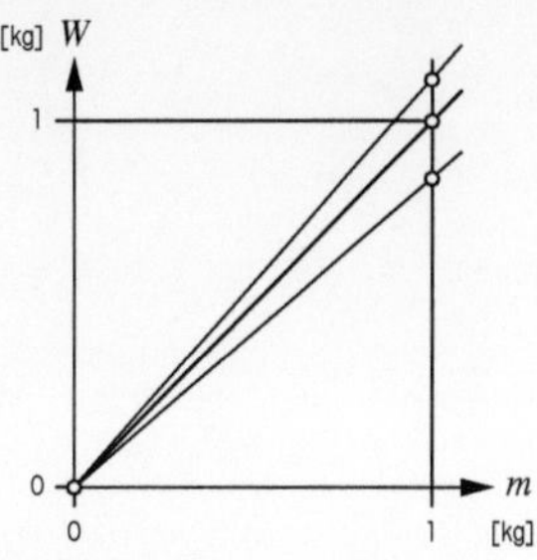

Fig. 149
Sensitivity between weighing value W and load m for a weighing instrument with a nominal range of 1 kg.
The middle line is the characteristic curve of a weighing instrument with correct sensitivity (slope). The upper line is too steep (sensitivity too high, shown exaggerated), the lower line too flat (sensitivity too low).

sensitivity adjustment

Operations for setting the →sensitivity of weighing instruments. To set the sensitivity, at least one →reference weight is placed on the instrument manually or by motor (→automatic adjustment). The weight is weighed and the measured value stored. The sensitivity of the weighing instrument is then corrected by the necessary amount. Depending on the design of the instrument, this correction can be made by mechanical or electrical means (e.g. adjusting screws or potentiometer). In the case of electronic weighing instruments, all subsequent weighment values are multiplied by a correction factor that is obtained by division of the nominal value of the reference weight by the corresponding measured value stored. If two or more reference weights are available, the →linearity can be adjusted in addition.

sensitivity drift

Change in the sensitivity caused by changing environmental effects, e.g. the →ambient temperature or heat dissipation from the electronics (→temperature drift, →switch-on drift), by fluctuations in air pressure (→air buoyancy), or through the passage of time (→long-term stability). →warm-up time, →drift, →automatic adjustment

sensitivity error

Obsolete term for →sensitivity offset.

sensitivity offset

1. Deviation of the →sensitivity from its true value. For →mechanical weighing instruments that do not function by the substitution method, the sensitivity mostly depends on the mass of the load. For electronic weighing instruments, the sensitivity depends on several factors that include, for example, the →leverage of the load cell (if present), the mechanical elasticity of the →spring ele-

ment of →strain-gage scales, the conversion constant of the →electrodynamic converter of →EMFC weighing instruments, as well as several electronic components such as resistance and/or voltage references. A too large or too small sensitivity causes measurement deviations that are proportional to the →net weight (→weighed-in quantity) (Fig. 149).

2. →Specification: Magnitude of deviation of the →sensitivity (measured between →zero load and →nominal load) from its correct value (=1) immediately after the sensitivity was adjusted (→adjust) with the built-in adjustment device, generally specified as a limit value (non-dimensional parameter).

sensor

Element of a →measuring instrument or →measuring chain upon which the →measurand acts directly and usually modifies an electrical output variable (→measurement signal), e.g. →strain gage.

serial data transfer

The consecutive transmission of data over one or several lines. →data transmission

set for the determination of density

→density determination set

set to zero

Bringing the indication of the unloaded →weighing instrument to zero. (Compare: →tare)

settling

A weighing instrument is in equilibrium (stable) when, after a change in load, all moving parts that are involved in the weighing process (unhindered by stops) have reached the →equilibrium position (stable position, →rest position).
For instruments that have a poorly damped oscillating lever system, the equilibrium position can be approximately calculated from the reversal points of the equilibrium indicator. In electronic weighing instruments, the equilibrium position is reached when the →measurement value becomes stable. This requires the load cell to have reached equilibrium and, if present, also the →signal filter. Printed or stored weighing results must not deviate by more than 1 →verification scale interval from the final weight value ([OIML R 76–1] T.4.4.2). →stand-still detector, →stand-still lock

settling position
The equilibrium position is that position of a movable measurement system in which there is →equilibrium between all
forces acting on the system. On balances with an →incli-
nation range, within this range any equilibrium position is
possible.

settling time
1. The time that elapses between placing the weighed object on a weighing instrument (the object touches the
   →load receptor) and indication of a sufficiently stabilized
   →weighing value. →weighing time, →integration time,
   →stand-still detector
2. →Specification: Settling time, usually stated as a typical
   value (taking into account the influence of environmental
   conditions, configuration of the weighing instrument and
   weighed object). Stated in [s]. →weighing time

shipping lock
A device designed to lock all delicate measuring components in position to protect them against damage during
transport of the weighing instrument.

SI units
→International System of Units

signal
Conveyor of information; in metrology, specifically a →mea-
surement value.

signal filter
→filter

signal processing
1. Processing, modifying, or extracting analog and/or digital
   information from a →measurement signal. Examples
   of units that process signals are electronic sensors for
   →strain gage load cells, →analog-digital converters,
   →filters, and →stand-still detectors.
2. Preparation of information for transmission from an information source to an information sink. →interface

signal processing unit
→Electronic assembly that processes signals. →signal
processing

significant
→metrologically relevant

single component weighing instrument

→Automatic gravimetric filling instrument that is used in mixing facilities for apportioning (→apportion) or →weighing and always delivers a preset amount of the same material in individual or repetitive weighing operations. In contrast to a →multicomponent weighing instrument, a single component weighing instrument is required for each component of the mixture. In statistical surveys, the term "single component weighing instrument" is used as a generic term for all →filling scales.

single point load cell

→spring element, Fig. 153a

Single Range (SR)

The (single) range of a weighing instrument (→normal range).

single range weighing instrument

Weighing instrument with (only) one weighing range (→normal range).

single-pan balance

A balance that has only one weighing pan. Virtually all modern balances have only one weighing pan. →multi-pan balance

single-range balance

In contrast to a →multiple range instrument or →multi-interval instrument
1. weighing instrument with a →weighing range, which possesses only one →scale interval and one maximum capacity;
2. verified weighing instrument with a →weighing range, which possesses only one →verification scale interval and one →maximum capacity.

sinker

A →displacement body of known volume, usually made of glass, that is suspended from a wire and immersed in a liquid to determine the →density of the liquid (Fig. 150a). Since the volume of the sinker is known, the density of the liquid can be determined directly from its →buoyancy (Fig. 150b) (→density determination). →plunger

skip scale

→Automatic weighing instrument →to weigh bulk goods, e.g. for disposal vehicles (→garbage scale). The →load

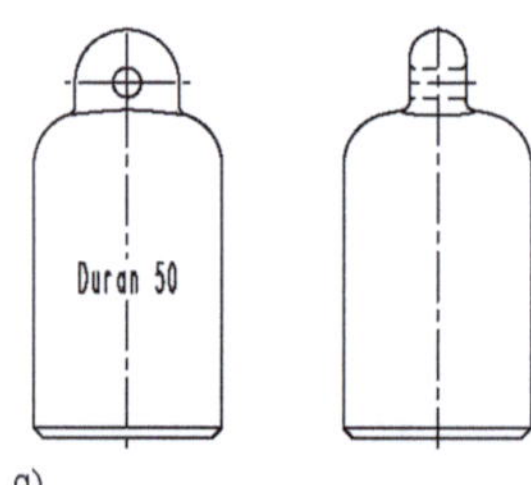

a)

b)

Fig. 150
Sinker
a) sinker made of glass;
b) using a sinker to determine the density of a liquid

cells are integrated into the handles, forks or grippers that lift the bulk goods container. Weighing takes place automatically during loading and/or unloading of the container. Should it not be possible to stop movement of the container for tare and gross weighing, acceleration sensors are used to compensate the dynamic forces. →vehicle on-board weighing system

Fig. 151
Skip scale
(Image by courtesy of Digisens AG, Murten, CH)

sliding weight balance

A lever balance (Fig. 152) in which →load compensation is effected by a →rider system (Fig. 122). The →riders are set by hand or by means of a setting device. The →equilibrium position is found by moving the →rider. The →Roman beam scale is an example of a simple sliding weight balance with →hanging pan. →physical weighing principle

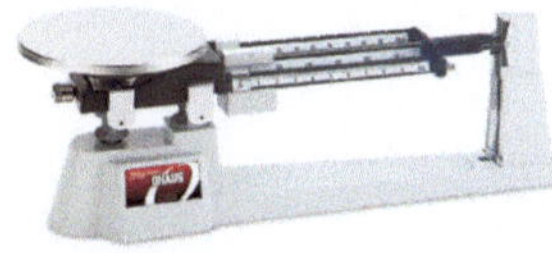

Fig. 152
Sliding weight balance with three sliding weights (riders or poise weights)

slope

→sensitivity

smallest acceptable amount

For →automatic weighing instruments and for →conveyor belt weighers, in applications subject to legal metrology the smallest acceptable amount of weighed material. →minimum capacity (compare: →minimum sample weight)

software

A term for all programs or non-fixed components that are not electronic or mechanical parts of a computer, as opposed to →hardware. A distinction is made between operating software and user software. Operating software includes, for instance, the program built-in by the computer manufacturer to start the hardware and load the operating system with its numerous commands and user programs. User software includes the control of working procedures, interaction with the user, and processing of the measurement data. →firmware, →weighing software

software identification
Sequence of readable characters of a software that is inseparably associated with this software (e.g. version number, checksum) ([OIML R 76-1] T.2.8.6).

software securing
Software means of sealing components or setting elements to which access or changes are forbidden. If access or changes have taken place, they must be apparent to the user ([OIML R 76-1] 4.1.2.4, 5.5.2.2).
Example: A non-resettable event counter within the →legally relevant software is automatically incremented every time there is an access or change. The counter can be indicated at any time at a keystroke and compared by the user with the reference value. The reference value for the counter at the time of verification is shown on a sealed plate (verification stickers). The weighing instrument may only be used in verified operation if the counter agrees with the reference value.

software separation
Unambiguous separation of software into →legally relevant software and non-legally relevant software. If no software separation exists, the whole software is to be considered as legally relevant ([OIML R 76-1] T.2.8.7).

software, legally relevant
→legally relevant software. ([OIML R 76-1] 2.8.1)

SOP
Abbreviation for →standard operating procedures.

sort, to
Separation of similar items and assignment to, for instance, →weight classes. (compare: →classify according to mass)

sorting balance
A limit weighing instrument with which similar items are sorted (→to sort) according to →weight classes.

specific weight
The specific weight γ of a body is the ratio of its →weight F_G to its →volume V

$$\gamma = \frac{F_G}{V}$$

In contrast to →density, specific weight depends on →gravity. Specific weight should not be confused with specific gravity, which is an obsolete term for relative density.

specification
Quantitative and qualitative declaration (→tolerance) that describes a property or characteristic (e.g. →repeatability, →nonlinearity) of a (measuring) instrument. Sometimes, a differentiation is made between guaranteed and typical specifications. The value of the characteristic that is measured on the individual instrument must usually conform to the guaranteed specification after installation and after maintenance work (→Equipment Qualification). A typical specification is based on the value attained by a large number of instruments. However, this value need not necessarily be matched by all instruments. With the assistance of the specification, an instrument's suitability for its intended use can be evaluated before it is purchased (→Design Qualification). A specification is usually stated as a (usually two-sided) tolerance interval (e.g. →nonlinearity), or as a standard deviation (e.g. →repeatability).
The most important specifications for describing the behavior of a →weighing instrument are the nominal characteristics →readability and →weighing capacity, and the metrological characteristics →repeatability, →eccentric load, →nonlinearity, →sensitivity, →temperature drift (of the sensitivity), and →stability (of the sensitivity). Further characteristics are →hysteresis, →zero point stability, →temperature drift of the zero point, →load drift.

spirit level
→level indicator

spring constant
Quotient c of a change in the force ΔF acting on a →spring element and the resulting change in the strain Δs (compression or extension)

$$c = \frac{\Delta F}{\Delta s}$$

spring element
Part of the load cell (e.g. →strain gage load cell, →spring scale) that is subjected to the weight force and is thereby deformed (Fig. 153). The critical property of the spring element is its elasticity, which is the link between the weight force being measured and the resulting deformation. Typical

spring elements are helical springs, compression column, twin flexible beams, shear flexible beams, circular springs, etc.

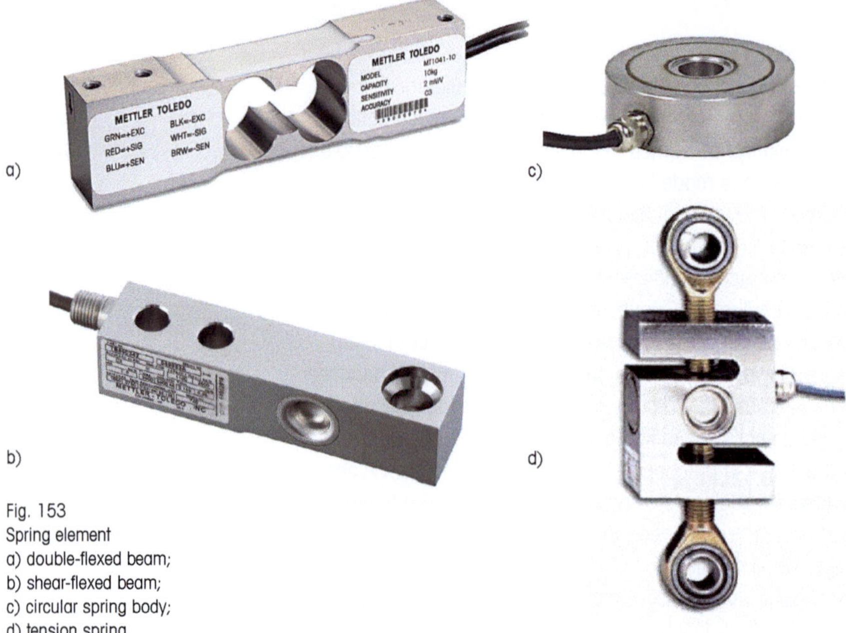

Fig. 153
Spring element
a) double-flexed beam;
b) shear-flexed beam;
c) circular spring body;
d) tension spring

spring force
Force that is exerted by an elastically deformed body on its environment.

spring measurement device
→spring element

spring scale
1. Generic term for weighing instruments in which measurement of the weight force uses the principle of elastic deformation of one or more springs (→spring element) (Fig. 154). Spring weighing instruments in which the load receptor is suspended directly from a spring without any transmission elements (levers, hydraulics) are referred to as simple spring scales.
2. A usually cylinder-shaped weighing instrument with a spiral spring (Fig. 155). The scale is hung by an eye and the lower hook is loaded with the item that is to be weighed.

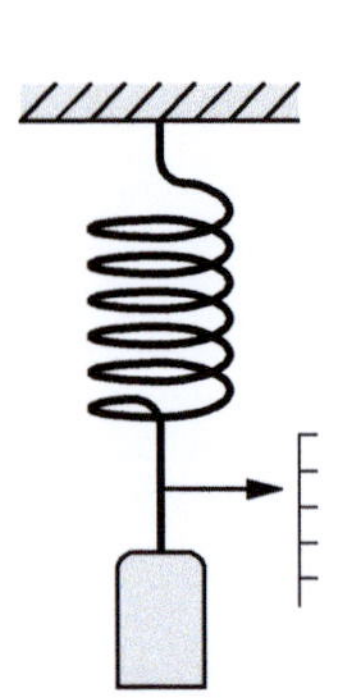

Fig. 154
Weighing principle of a spring scale, here with helical spring

SQC
Abbreviation for →statistical quality control.

stability

1. Mechanical stability of a weighing instrument. →stability test
2. Constancy of a metrological property (→specification) over time.

stability of the sensitivity

1. The amount of the change in sensitivity over time.
2. →Specification: Magnitude of difference in →sensitivity between two →adjustments with the built-in adjustment device, always measured immediately after the respective adjustment, relative to the intervening time interval. Usually expressed as a limit value in [1/a] (per annum, per year).
2.1 If the measuring instrument has no built-in adjustment device, the maximum difference in the sensitivity between two →adjustments with an external reference mass is meant, ignoring the effect of the latter.

stability test

Testing of a weighing instrument for mechanical stability in which certain →test loads are placed on the →load receptor lengthwise as well as crosswise.

stabilization time

→settling time

stamping label

A seal holder that carries the →verification stamp or →securing sticker. The stamping label must be securely fastened to the weighing instrument.

stamping mark

Stamping marks are marks that are applied by the manufacturer or the →Weights and Measures authorities to instruments that are subject to legal metrology requirements (→verification). The stamping mark confirms the conformity of the respective instrument with the underlying laws and/or regulations (→legal metrology requirements). →EC verification marking, →main verification mark

stand-still

The state of an equilibrated weighing instrument with stable →indication. →settling, →final weight value

stand-still detector

Device that monitors the attenuation of the →settling and

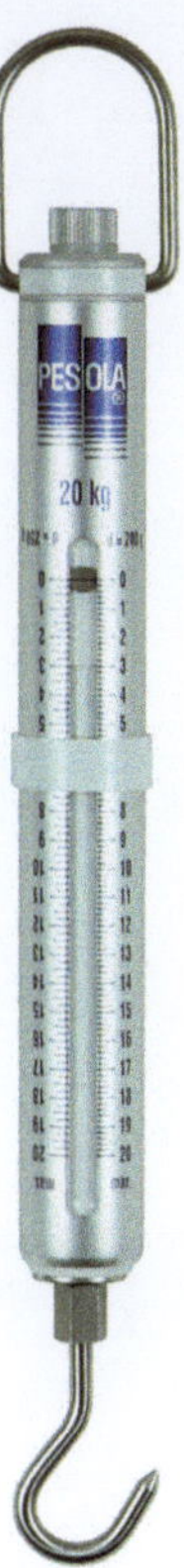

Fig. 155
Spring scale
(Image by courtesy of Pesola AG, Baar, CH)

deduces from it the →stand-still of a weighing instrument (→final weight value). The determination of the instant is always an estimate, since the stable, true measurement value is unknown in advance.

stand-still lock

A device, installed primarily on weighing instruments equipped with a printer and computer, which ensures that a →weighing result is not printed or forwarded to the PC until the measurement value has attained →stand-still. →settling

standard

1. Realization of the definition of a given quantity, with stated quantity value and associated measurement uncertainty, used as a reference. ([VIM:2008] 5.1)
2. Comparison object (material measure) or accurate measuring instrument that is used to →calibrate less accurate standards. →mass standard
3. →Weight piece of constant and known mass.

standard deviation

In probability theory and statistics, a measure of the value of the mean (quadratic) distribution of a measurand about its →mean value. The empirical dispersion s_x can be estimated from a sample $\{x_i\}$ of n values as follows

$$s_x = \sqrt{\frac{1}{n-1}\sum_{i=1}^{n}(x_i - \bar{x})^2} \quad \text{where} \quad \bar{x} = \frac{1}{n}\sum_{i=1}^{n} x_i$$

→normal distribution

standard gravity

A nominal value for →gravity at sea level and approximately 45° latitude, defined by convention in 1901 to be

$$g_N = 9.80665 \text{ N/kg} \quad ^{16}$$

(Fig. 77). This value lies within the range of gravity variation. →local gravity, →Bouguer anomaly

standard load

In field of legal metrology: A load consisting of certified or verified →weight pieces, weighing equipment, or →standard test vehicle.

standard measurement uncertainty

→standard uncertainty

---

[16] 1 N/kg = 1 m/s^2

standard operating procedures

In the context of →Good Laboratory Practice (GLP), standard operating procedures (SOP) are written directions for the performance of certain constantly recurring laboratory investigations or other activities that are usually not described in greater detail in test plans or test guidelines.

standard range

→normal range

standard test package

A package used to determine the tolerance limits of →checkweighers. The standard test package must satisfy certain conditions with respect to mass, dimensions, and material, that are stipulated in the →Measuring Instruments Directive.

standard test vehicle

1. In the broader context: Vehicle used to transport →standard weights for the →verification of →weighing instruments with relatively high maximum capacities (e.g. rail or road vehicle scales) (Fig. 156).
2. Strictly by definition: Vehicle according to 1. that has a certain weight of its own that serves as the →standard load and that can usually also transport additional →standard weights.

Fig. 156
Standard test vehicle
Verification of a bridge scale with standard weights.
(Image by courtesy of Grimm Waagen, Tresdorf bei Korneuburg, AT)

standard uncertainty

Measurement uncertainty expressed as a →standard deviation. ([VIM:2008] 2.30)

standard weight

Short form for →standard weight piece.

standard weight piece

A →weight piece used especially by departments of weights and measures to test, →calibrate, or →adjust other weights or weighing instruments. The accuracy of the →standard weight is higher (usually by a factor of at least three) than the object being tested. →hierarchy of mass standards and weights

standby operation

A special operating mode of a →measuring instrument: When the instrument is in standby mode, only the display is shut down, all other electronics (e.g. →load cell and evaluation electronics of a →weighing instrument) remain in operation, e.g. to avoid →switch-on drift.

statistical confidence

The probability with which a given number of measurement values can be expected to fall within a given range (→confidence level). →coverage interval, →measurement uncertainty

statistical quality control

Collective name for all of the measures involved in manufacturing processes for monitoring, checking, and optimizing filling processes (→prepackage), a.k.a. →SQC. Statistical quality control is used to monitor all aspects of the manufacturing process to ensure that the required specifications, as well as the economic and legal requirements (→Prepackaged Products Directive), are fulfilled. Statistical values (e.g. →fill quantities) that are obtained from →samples provide continuous information on the status of the production process. The production process can be adjusted if necessary, or in exceptional situations it can be stopped. →obligation to record, →prepackage process control, →filling process control facility, →sample

Fig. 157
Checking the weight of samples of a pharmaceutical product. Left checkweigher, middle capsule feeder.

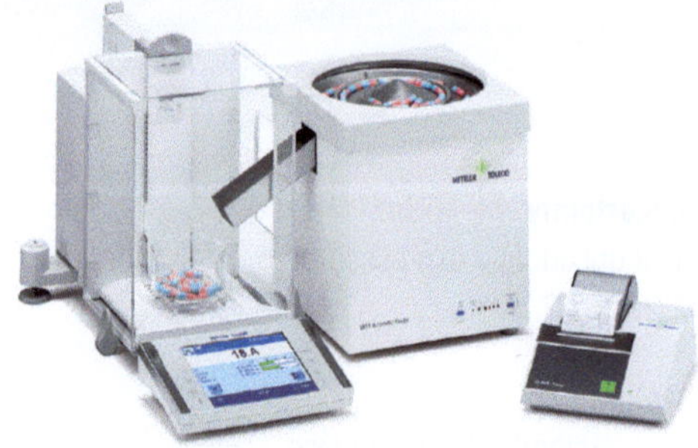

statistics

In connection with →weighing, the statistical evaluation of weighing results. →application module

step method

A method which, under certain verification conditions, can be used to test the →trueness of a weighing instrument with a relatively high maximum capacity, using a load that may not be fully known. The step method may be complete or abbreviated. The latter is used only for a few models of weighing instruments, and only if the →weighing-out device, such as the sliding weight beam, has already been tested.
→verification procedure for weighing instruments

strain gage

Electrical resistance element consisting of an electrically conductive, meandering foil trace bonded to a strip of non-conductive carrier film (Fig. 158a). This carrier film is in turn bonded to the structure whose strain (elongation) is to be measured. Typical resistances of strain gages range from 30 to 5000 Ω, with 120 Ω, 350 Ω and 1000 Ω being the most common values. When the strain gage is stretched, the length of the electrical conductor increases while its cross-section decreases. Both of these effects increase the electrical resistance of the conductor (Fig. 158b). Strain gages are subjected to either tensile or compressive strain of a few percent, in special cases up to 20%. To preserve transducer →linearity, the strain of the gages used on →strain gage load cells is limited to around 0.1%. With a 350 Ω metallic gage whose →gage factor is about 2, this yields a resistance change of 0.7 Ω. Depending on the application, strain gages with different conductor patterns are used (Fig. 159) to suit the particular strain field. The active length of a strain gage typically ranges from 0.2 to 100 mm. Strain gages are commonly connected in a →Wheatstone bridge to accurately measure their small changes in resistance.

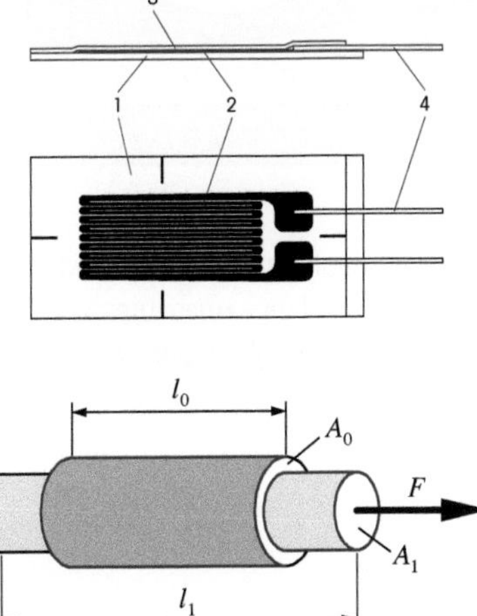

Fig. 158a
Diagrammatic structure
of a strain gage

1: non-conductive carrier film
2: electrically conductive measurement grid (meander)
3: non-conductive encapsulation film
4: electrical connections

Fig. 158b
Principle of the changing resistance of an electrical conductor:
Cross section A and length l of the conductor when relaxed (suffix 0), and under the tensile force F (suffix 1)

Fig. 159
Conductor structures (meanders) for different applications

a) Unidimensional structure for measuring linear strain (general use, and for conventional load cells);
b) T-rosette for measuring direct and Poisson strain (use in so-called pin cells, Fig. 132)
c) orthogonal structure offset at 45° to the principle axes for measuring shear strain (used in so-called shear beam cells, Fig. 153b);
d) round rosette for measuring the strain of round diaphragms (used in pressure sensors).
(Images by courtesy of Vishay Micro-Measurements, Raleigh (NC), USA)

a)

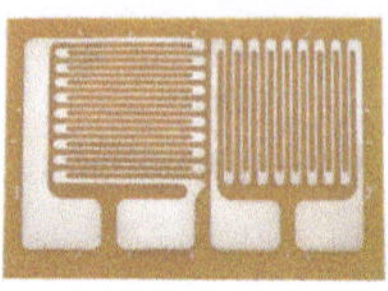
b)

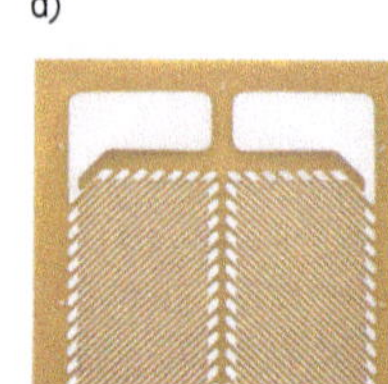
c)

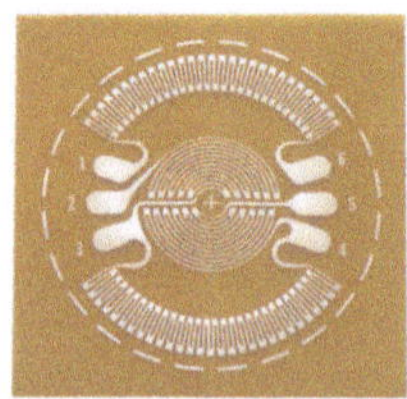
d)

strain gage load cell

Load cell in which the applied →weight force causes an elastic deformation of a →spring element, which is usually metallic. The deformation is sensed by means of →strain gages (Fig. 160). →physical weighing principle, →converter

Fig. 160
Diagrammatic structure of a strain gage load cell with two tension and two compression strain gages

1: spring element
2: strain gages
3: tension zone
4: compression zone
F_G: loading by the weight force

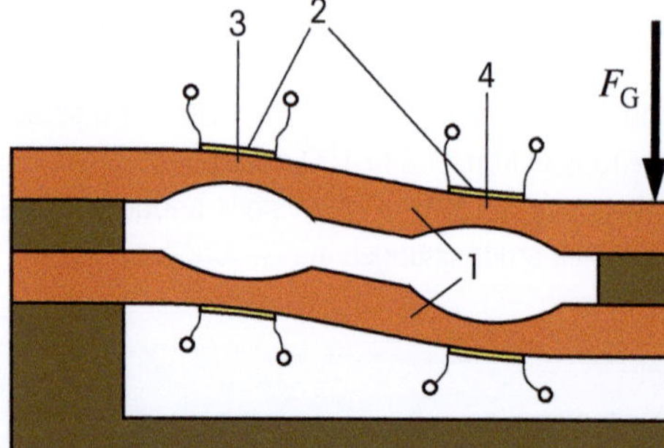

strain gage scale

An →electromechanical weighing instrument in which a →strain gage load cell is used as →measurement transducer (Fig. 161). →physical weighing principle

strain gauge

→strain gage

strain gauge load cell

→strain gage load cell

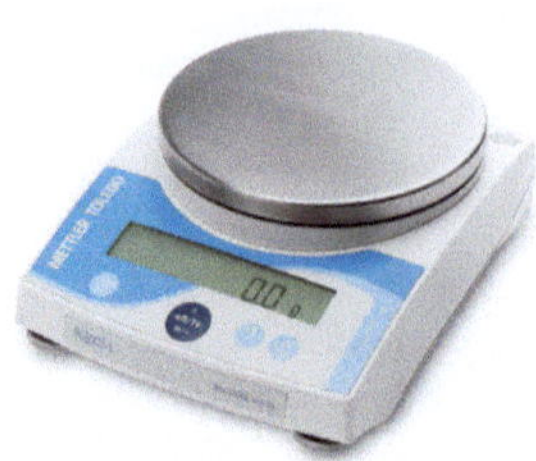
Fig. 161
Strain gage scale

string

Thin, taut cord, usually made of metal, that is capable of vibrating. The frequency of vibration (resonant frequency) of a string depends (nonlinearly) on its tension. →string balance

string balance

An →electromechanical weighing instrument in which a →string load cell is used as a measurement →converter. →physical weighing principle

string load cell

Load cell in which the weight force of the weighed object modulates the tension force of a →string (Fig. 162). The string is excited to vibration by an electrodynamic converter. An increase in load increases the resonant frequency of the string. The change in frequency is a measure of the weight force that is to be determined. The frequency of vibration can be registered as a digital quantity. →physical weighing principle, →converter

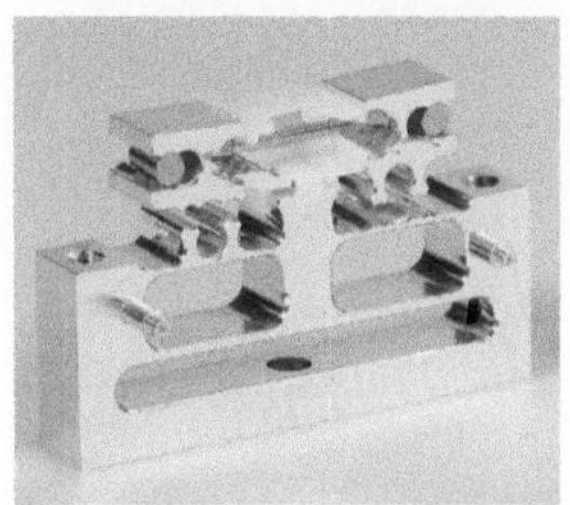

Fig. 162a
String load cell

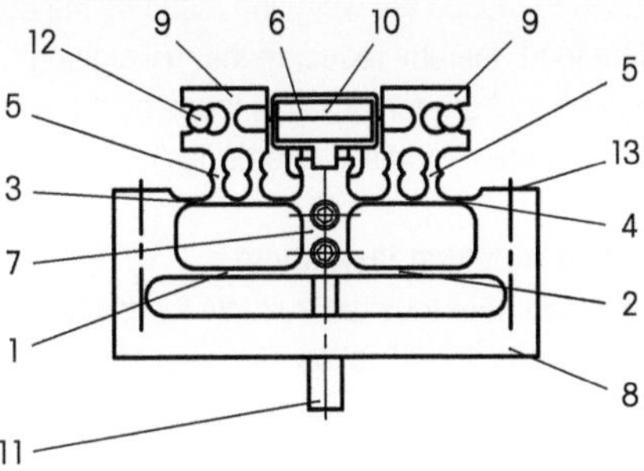

Fig. 162b
Diagrammatic cross section of a
string load cell

1...4: double parallelogram
5: spring coupling
6: vibrating string
7: central block
8: body base
9: clamping block
10: permanent magnet
11: force transmission
12: tensioning pin
13: fastening surface

(Images 162a and b by courtesy of
Digisens AG, Murten, CH)

subsequent verification

Any →verification of a measuring instrument after a previous verification, including mandatory periodic verification (before or after expiration of the →validity period of verification) or verification after repair, contrary to →initial verification. ([VIML] 2.16).

substitution balance

Balance in which the weighing sample and the →weight pieces are on the same end of the lever arm (Fig. 163). With no load on the pan, the weights and the constant counterweight are in →equilibrium. With the weighing sample on the pan, the corresponding mass is removed from the weight set (→dial weights) to re-establish the equilibrium (Fig. 124). By comparison with the two-arm balance, the substitution principle (→physical weighing principle) has the advantages that

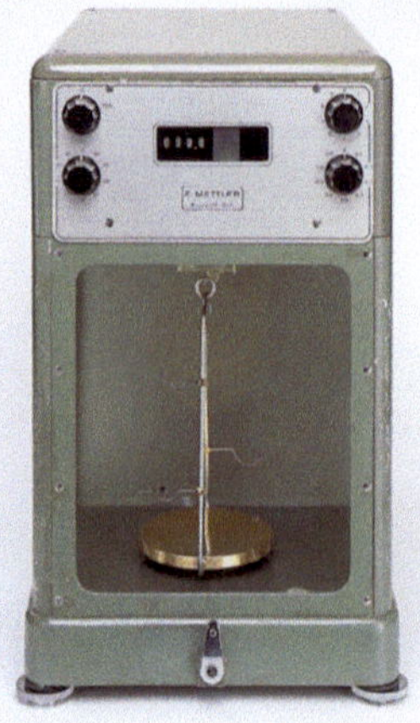

Fig. 163
Traditional substitution balance

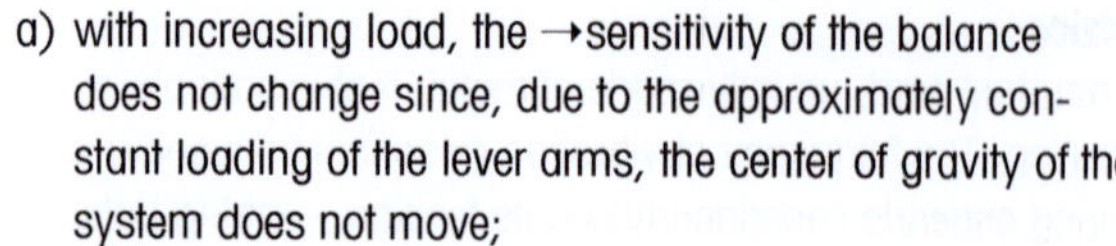

a) with increasing load, the →sensitivity of the balance does not change since, due to the approximately constant loading of the lever arms, the center of gravity of the system does not move;
b) the sensitivity is independent of the nominal value of the →mechanical advantage of the lever, since the weighing sample and the weight pieces are compared on the same side of the lever (→Borda weighing method; compare: →Gaussian weighing method)

substitution weighing

A method of weighing (→Borda weighing method) in which first the sample and then the weights of the same mass value are compared with one and the same auxiliary load (called the →tare load). →substitution balance, →physical weighing principle. (compare: →Gaussian weighing method)

subtractive tare device

A device used to reduce the weighing result by the amount of the →tare load, thereby reducing the →weighing range of the weighing instrument for →net loads by the same amount. (compare: →additive tare device)

suitability of a weighing instrument

A weighing instrument must be suitable for its intended purpose, for use, and for verification. The weighing instrument must be constructed in such manner that, for example,
– it fulfils the requirements of the respective application and environmental conditions;
– its metrological characteristics remain constant for a specified period of use;
– verification tests can be performed;
– the standard weights can be easily and safely placed on the load receptor.
→Directive on Non-Automatic Weighing Instruments, →Measuring Instruments Directive, →EN 45501

support

→Weighing table, console, or frame on which the weighing instrument is installed and used.

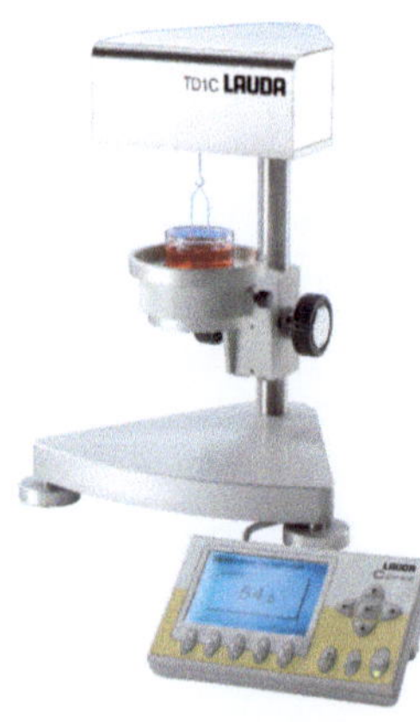

Fig. 164a
Balance for measuring surface tension
(Image 161a courtesy of LAUDA Dr. R. Wobser GmbH & CO. KG, Lauda Königshofen, DE)

surface tension

A property of the surface at the boundary between a liquid and a gas, e.g. between water and air. The surface of a liquid behaves like a taut elastic film. The unit of surface tension is N/m.

surface tension balance

Balance for measuring the tension that the surface of a liquid exerts (→surface tension) (Fig. 164a). For this purpose, the force exerted by the liquid on an immersed wire stirrup (Lenard's method, Fig. 164b), wire ring (De Noüy, Fig. 164c), or plate (Wilhelmy, Fig. 164d) is measured when it is withdrawn from the liquid.

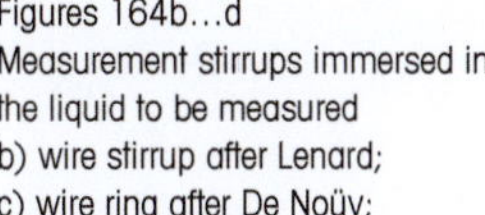
b)

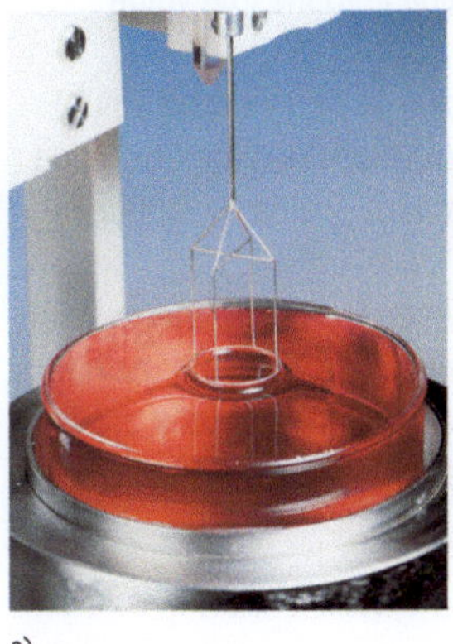
c)

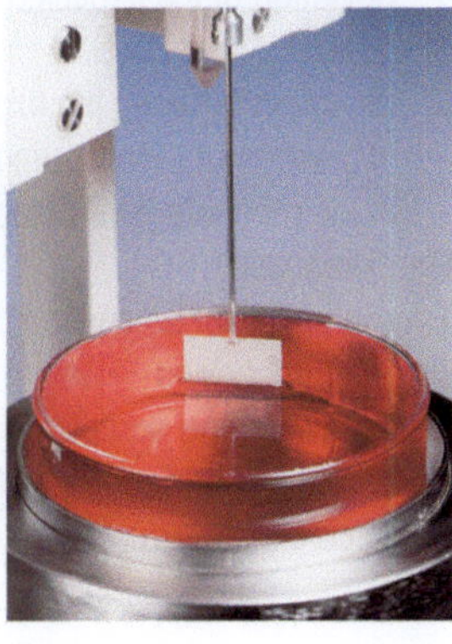
d)

Figures 164b…d
Measurement stirrups immersed in the liquid to be measured
b) wire stirrup after Lenard;
c) wire ring after De Noüy;
d) plate after Wilhelmy

(Image 164b from Wikimedia Commons (Author: Michael Krahe) is available under the GNU license [17]; images 164c and 164d courtesy of LAUDA Dr. R. Wobser GmbH & CO. KG, Lauda Königshofen, DE)

switch-on behavior

→warm-up time, →drift, →switch-on drift

switch-on drift

Measurement value drift (→drift) caused mainly by heat dissipation from the electronics of a weighing instrument when it is put into operation (→zero point drift or →sensitivity drift).

switchoff criterion

Criterion of a →dryer that determines when drying of a sample will be terminated. Drying can be terminated, for example, when the decrease in weight per unit of time falls below a specified value.

---

[17] GNU Free Documentation License: http://www.gnu.org/licenses/fdl.txt

system scale

A →scale which is part of a system of scales with or without
a switching arrangement in which various →weighbridges
(→load receptors) are, or can be, jointly connected to one
→weighing-out device. The →display device generally includes not only the →weighing results for each individual
weighbridge (individual load) but also the weighing results
for all weighbridges together (total load).

systematic deviation

→systematic error

systematic error

1. Component of measurement error that in replicate measurements remains constant or varies in a predictable
   manner. ([VIM:2008] 2.17)
2. Deviation of the expected value from the true value.
   ([DIN 1319-1] 3.5.2)
3. Deviation between the expected value (→mean value)
   of a series of →measurement values and the true value
   of the measurement object, a.k.a. →bias. (Compare:
   →trueness)

Example: Deviation caused by imperfections in the →weigh-
ing instrument and →weight pieces, the weighing method,
and the weighing sample as well as metrologically detect-
able →influence quantities. The deviation must be corrected
by calculation. →buoyancy

t
Unit symbol for the mass unit →metric ton.

T
1. Symbol for →tare value.
2. Customary unit symbol for the (non-metric) mass units
   of the "short ton" and "long ton", to keep them distinct
   from the →metric ton ("t").

tael
→Nonmetric unit of mass for precious metals (unit symbol
"tl"), used in Eastern Asia. The tael is defined differently in
different countries:

Hong Kong tael	1 tl ≈ 37.429 g
Singapore tael	1 tl ≈ 37.79936 g
Taiwan tael	1 tl = 37.5 g

tank scale
→Scale for tanks, usually with weighing cells between
each of the tank legs and the foundation, for →weighing or
apportioning (→apportion) →fluids.

tare
1. That part of a weighing sample that is not the object of
   the weighment, but which cannot be separated from the
   actual load, such as, for example, a container (e.g. crate,
   bottle), a transportation device (e.g. pallet), or packaging.
2. Non-technical term for the mass of the tare (→tare
   weight).

Fig. 165
Tank scale
Between the tank legs and the
foundation the built-in load cells are
visible.

tare compensation device
A device used to compensate a →tare load without indicating the tare value on the loaded weighing instrument.

tare device
Generic term for →tare compensation device and →tare
weighing device ([OIML R 76-1] T.2.7.4). A further distinction is made between →additive tare devices and →sub-
tractive tare devices, as well as between automatic, semi-
automatic, and non-automatic tare devices.

tare load
1. Packaging, transport container, or vessel in which the
   sample is weighed. →tare weight
2. Load that is not the object of the weighing, but is required
   to determine an unknown mass, e.g. as auxiliary or

compensating load in the →Borda weighing method or →substitution weighing method. It may also be a weight that is fastened on the →load receptor or →lever.

tare memory
A feature of an →electromechanical weighing instrument which makes it possible to store and then recall the →tare weight.

tare signal
Signal (e.g. lamp, or character in the display) which indicates that the →tare device is being used.

tare value
Weight value of a load, determined by a tare weighing device, often designated with symbol T ([OIML R 76-1] T.5.2.3). →tare weight

tare weighing device
Device for →weighing a →tare load that allows the tare value to be indicated or printed with the weighing instrument either loaded or unloaded ([OIML R 76-1] T.2.7.4.2).

tare weight
Weight of the →tare that is weighed with the sample; empty weight. →gross weight, →net weight

tare, to
To compensate the →tare load with or without determination of the tare load. (In non-technical language, incorrectly used for →set to zero.)

target fill quantity
→Fill quantity that a package should contain for the package to comply with the law. The target fill quantity is generally the fill quantity that is added in the filling process. The target fill quantity is composed of a →nominal fill quantity and an overfill.

target value
For given conditions, a value specified between the maximum and minimum permissible values.

taring material
A material such as steel shot which is added to the →adjusting cavity to →adjust a →weight piece to the →nominal value or to bring the →display device of an unloaded weighing instrument to zero.

taring range
A range within which the indication of a weighing instrument
can, or is permitted to, be set to zero by means of the →tare
device.

taut band suspension
Support of, for instance, the →balance beam by means of
a taut metal band that is at right angles to the pivotal plane.
→flexible coupling

temperature compensation
A device or measure used to compensate the effects of
changes in temperature on the →measurement value. The
effects can be compensated by mechanical means (e.g. a
special arrangement of springs or choice of materials), by
electronic means (e.g. measuring the temperature with a
→sensor and analog compensation), or by computer-aided
signal processing using an algorithm. The effects of temperature are determined by →weighing →reference weights
while the weighing instrument is systematically exposed to
different ambient temperatures.

temperature drift
Slow change over time in the value of a metrological characteristic (e.g. of the measurement value) of a measuring
instrument with changing ambient temperature.
1. Temperature drift of the zero point
   Drift of the →measurement value of the unloaded weighing instrument with changing ambient temperature;
2. Temperature drift of the sensitivity
   Drift of the →sensitivity (→net value) of a measuring
   instrument with changing ambient temperature;
3. →Specification: Temperature drift of the sensitivity
   Magnitude of sensitivity drift of a measuring instrument
   with changing ambient temperature; usually expressed
   as a limit value, in [1/°C] or [1/K].
(Compare: →switch-on drift)

temperature influence
The →ambient temperature influences measurements in
a variety of ways. In the case of →weighing instruments
and →load cells, temperature changes cause expansion
(or contraction) of components (e.g. →levers, →flexible
joints, →force links), or changes in material properties
(e.g. →spring constants of →spring elements, magnetic flux
of →electrodynamic converters). →Weighed objects that
are not acclimatized to the →ambient temperature cause

transient effects (e.g. air currents). Particularly in the case of →high-resolution weighments, these effects can invalidate the result, either by causing →systematic errors, or causing the measured value to drift (→temperature drift), or causing the →repeatability to be impaired. Although these deviations can be compensated to some extent (→temperature compensation), →temperature limits are specified for the operation of instruments.

temperature limits
Limit values of the →temperature range. Usual temperature limits are -10°C to +40°C. Special temperature limits can be selected according to the purpose for which the weighing instrument is used, with the following minimum ranges:
5°C for weighing instruments of →accuracy class Ⓘ (→weighing instrument of special accuracy);
15°C for weighing instruments of accuracy class Ⓤ (→weighing instrument of high accuracy);
30°C for weighing instruments of accuracy class Ⓤ (→weighing instrument of medium accuracy) and Ⓤ (→weighing instrument of ordinary accuracy).

temperature range
Range of the →ambient temperature between the lower and upper →temperature limit within which a weighing instrument may be used. ([OIML R 76-1] 3.9.2)

tendency correction device
A device on →checkweighers that evaluates the weighing results and uses them to control an upstream filling machine to correct any tendency of the mean value of the added mass to shift.

tension weighing cell
→spring element, Fig. 153d

tensitometer
An instrument measuring →surface tension. →surface tension balance

terminal
Digital device with one or more keys to operate the weighing instrument and a display for the weighing results that are digitally transmitted from one or more →weigh modules or from an →analog signal and data processing device ([OIML R 76-1] T.2.2.5).

test

1. Generic term for testing a single function or a complete instrument.
2. Term used in non-technical language for →calibration.
3. Determination of performance and capabilities according to specified requirements, e.g. manufacturing test according to internal company standards, →metrological test, type approval test for a weighing instrument according to →Directive on Non-Automatic Weighing Instruments.

test certificate

An auxiliary document issued by a →Notified Body that serves to facilitate a type test. Although the document is issued for →modules or →auxiliary devices of →weighing instruments, it does not replace a →type approval. For a test certificate to be issued, the respective modules and auxiliary devices must fulfill the most important stipulations of the applicable guidelines and directives. Test certificates are only meaningful in association with a type approval. Depending on how a test certificate is worded in a type approval, it may allow the connection of an auxiliary device to, or integration of a module into, a weighing instrument.

test load

A load used to check a weighing instrument.

test report

Document of a →Notified Body, or a testing laboratory that is accredited (→accreditation) according to EN 17025, in which a test of a →module or →auxiliary device is described along with all metrologically relevant characteristics and particularities.

For modules or auxiliary devices that do not completely conform to →European Standard EN 45501 and the corresponding →WELMEC guidelines, but have been tested against the most important stipulations, test reports can be issued to facilitate the work involved in the type approval process (→type approval). Test reports may not be cited in a type approval in the sense of a →general clause.

test weight

A reference mass in the form of a →weight piece. The weight has the appropriate accuracy (→accuracy classes of weight pieces) and is used for regular checks of the functioning and trueness of the weighing instrument by the user.

testing mark
Generic term for →verification mark, →year mark, →year notation, →main verification mark, or →stamping mark.

tex
Unit of measure for the →yarn count 1 tex = 1 g/km

TGA
Abbreviation for →thermogravimetric analysis.

thermal analysis
→thermogravimetry

thermal printer
Printer that, by means of targeted activation or heating of heating points that are mounted on a strip produces visually recognizable characters on thermosensitive paper. The lifetime of such printouts is generally limited.

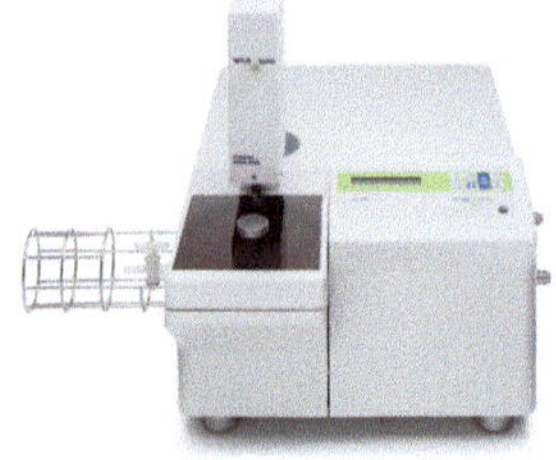
Fig. 166a
Thermobalance

thermobalance
A special balance with an oven to determine mass changes that may occur when the weighing sample is heated during the weighing process (Fig. 166). The balance has an automatic temperature control system as well as various devices that allow →weighments in special gases, air, or vacuum. For simultaneous investigation and identification of the decomposition products that are evolved, a thermobalance is often linked to an instrument such as a mass spectrometer or infrared spectrometer for analysis of the gas, as well as investigation and identification of the decomposition products that are evolved.

Fig. 166b
Cross section through a thermobalance:
at left, sample holder in oven;
right, the balance.

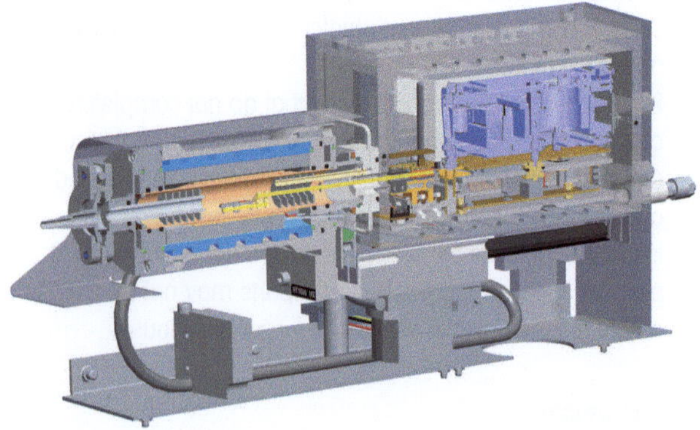

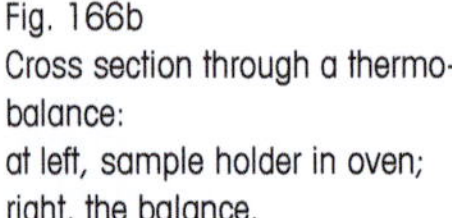

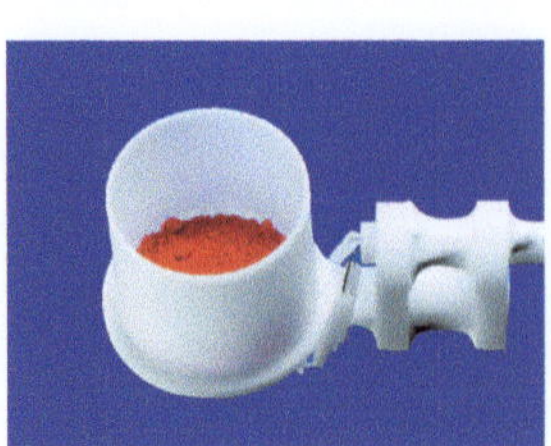
Fig. 166c
Sample holder for a thermobalance

thermogrammetry
1. Method of creating thermal images
2. Synonym for →thermogravimetry.

thermogravimeter

→thermobalance

thermogravimetric analysis (TGA)

→thermogravimetry

thermogravimetry

Thermogravimetric analysis (→gravimetry) is a quantitative and reproducible thermoanalytical method of measurement that allows the changing mass of a sample in relation to time and temperature to be determined with great accuracy (Fig. 167). The analysis of the material takes place under a well-defined reactive or inert atmosphere at up to high temperatures. The temperature-dependent processes that take place, e.g. vaporization, sublimation, or decomposition (chemical reaction), allow statements to be made about the thermal stability and decomposition characteristics of a material. →thermobalance, →dryer

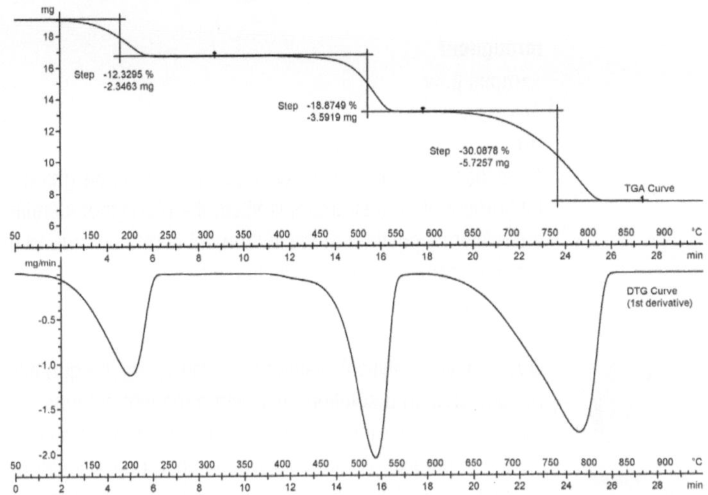

Fig. 167
Example of a thermogravimetric analysis, here for calcium oxalate monohydrate.
Upper curve: TGA curve (mass against temperature); lower curve: first derivative by temperature (change in mass against temperature)

three-knife balance

→Equal-arm beam balance with a total of three →knife-edge bearings (one in the middle and one at each end of the →balance beam) (Fig. 168). In the classic symmetrical beam balance, the known mass m_k and the unknown mass m_x cause mutually opposing moments of rotation. Since the two moments of rotation are rarely of equal magnitude, the beam usually comes to rest in an inclined position. The unknown mass m_x is given by the known mass m_k plus the supplementary mass m_a that corresponds to the angle of inclination α (→inclination range),

$$m_x = m_k + m_a$$

For a nonsymmetrical beam balance, the →lever ratio l_k/l_x must be taken into account. The following then applies:

$$m_x = m_k \frac{l_k}{l_x} + m_a$$

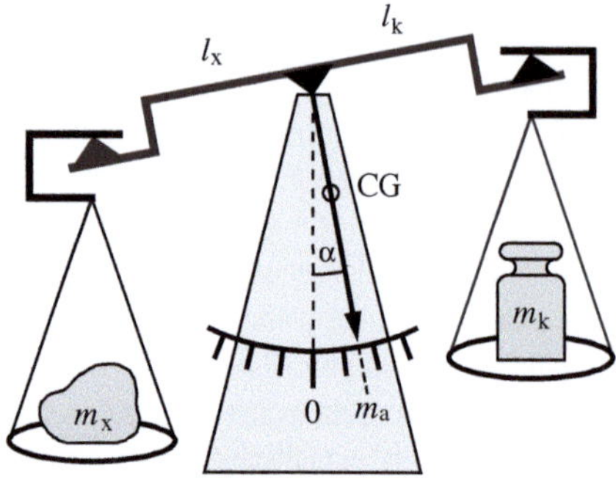

Fig. 168
Principle of the three-knife balance

m_x: unknown mass
m_k: known mass
m_a: supplementary mass corresponding to angle of inclination α
l_x: lever arm of the unknown mass
l_k: lever arm of the known mass
CG: center of gravity of the lever
α: inclination of the lever

through-balance hanger
→around-balance hanger

throughput
→mass flow

tilt
A deviation from the →reference position of the weighing instrument. If an instrument is tilted, the plane that identifies the reference position of the instrument is rotated around an arbitrary horizontal axis. →inclination

titration
Quantitative analytical method for determining the quantity of substance in a solution, in which a reagent of known concentration (titrant or standard solution) is dispensed from a →burette into an unknown solution until equilibrium is reached. Equilibrium is detected by, for example, the change in color of an indicator or, in the case of acid-alkali titrations, by measurement of the pH value. The volume (→volumetry) of standard solution required allows the amount of substance in the unknown solution to be calculated (Fig. 169).

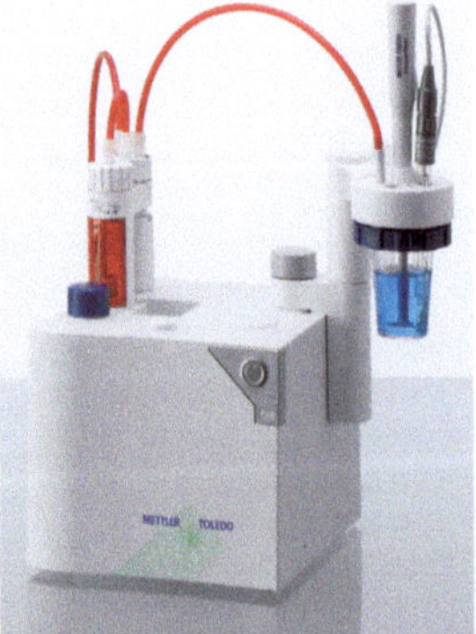

Fig. 169
Titrator

tolerance
The difference between the permissible maximum and minimum values of a →nominal value.

tolerance limit
A limit that is defined by the →tolerance about a specified →nominal value , which must not be exceed or fallen be-

low. Depending on the situation, the tolerance may be used one-sidedly or two-sidedly. →specification, →prepackage process control.

top-loading

1. Designates the type of design of a weighing instrument in which the →load receptor is located above the →balance beam, the →lever, or the →load cell. In contrast to a →low-level pan, the load receptor of a top-loading weighing instrument is guided mechanically to prevent it from tipping, for example with the aid of a →parallel guide (Fig. 20, 98, 108, 116), and a top-loading load receptor does not oscillate. →Roberval scale, →Béranger scale
2. In the broader context: Designation for 'guided load receptor', 'guided pan'.

top-loading load receptor

→top-loading

torque balance

1. Measuring instrument that uses the principle of a →weighing instrument to determine the torque of, for instance, engines and machine tools. →torsion balance
2. Measuring instrument where the load applied generates a torque that is used to indicate the weight. →spring scale

torsion balance

A type of balance (Fig. 170) that was originally constructed by John Michell (*1724–†1793), and subsequently developed further by Henry Cavendish (*1731–†1810), who used it to determine the gravitation constant (→gravitation).

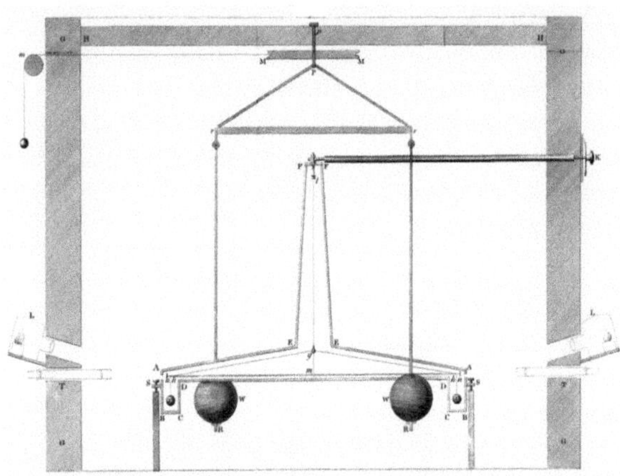

Fig. 170a
Illustration of the torsion balance
from Cavendish's original manuscript
of 1798

Fig. 170b
Experimental apparatus based on the principle of the torsion balance to test the →equivalence principle. Of the eight sample masses mounted on the pendulum body, four are of beryllium and four of titanium.
(Image by courtesy of the EötWash Group, University of Washington, Seattle (Washington), USA)

The torsion balance can be used to determine the mutual attraction of bodies (→gravitation) or the electrostatic force (→electric charge) between bodies. A torsion balance consists of a pendulum body that is suspended on a thin fiber. An external torque acting on the pendulum body causes the latter to rotate about the axis of the fiber until the opposing torque caused by the torsion of the fiber equals the external torque in magnitude. As a close approximation, the final angle of rotation, which is usually measured optically, is proportional to the external torque.

total control

Filling control in which all packages of a lot are checked. →prepackage process control

traceability

Property of a measurement result, or value of a →standard, to be related to suitable other standards, usually international or national standards, through an unbroken chain of comparison measurements ([VIM:2008] 2.41). The →standard weights that are used for mass determinations must always be traceable to the higher-level normals (Fig. 80).

tracing

Performance of the activities required to attain →traceability.

triangular support

Polygonal metal frame with small surface (→influence of moisture) that prevents round-bottom flasks or similar laboratory vessels that cannot stand alone from tipping over while being weighed on the weighing pan (Fig. 171).

triboelectricity

Physical effect in which contact and subsequent separation of two objects results in a residual surplus electric charge on one of the objects and a corresponding deficit of electric charge on the other object (→electric charge). Rubbing the objects together has the same effect. →electric charge, →influence quantities

triple-beam balance

→sliding weight balance

truck scale

Common term in the United States of America for →'vehicle scale'.

Fig. 171
Triangular support

trueness
1. Qualitative term describing the →systematic error of
   measurements.
2. The closeness of agreement between the expected
   value (→mean value) of a series of →measurement
   values and the true value of the measurement object
   ([ISO 5725] 3.7) (Fig. 1). Example: The ability of a
   measuring instrument to provide measurement values
   that coincide with the true value of the measured object.
   →accuracy (compare: →precision)

Note: The trueness can only be determined when multiple
measurement values, as well as a reference value that is
recognized as true, are available.

two-knife balance
A single-lever balance with a main knife-edge and a second
→knife-edge that supports both the →load and the →weight
pieces. At the opposite end of the lever from the second
knife-edge is a fixed counterweight.
→substitution balance, →design and function of a
mechanical balance

type approval
Decision of legal relevance, based on the evaluation report,
that the type of a measuring instrument complies with the
relevant statutory requirements and is suitable for use in
the regulated area (→legal metrology) in such a way that it
is expected to provide reliable measurement results over a
defined period of time ([VIML] 2.6). →EC type approval

type approval certificate
Document issued by the →Notified Body for the approved
design with technical details, restrictions, etc.

type evaluation
→type examination

type examination
An evaluation of performance, operating characteristics,
features and options, in which a →Notified Body tests and
certifies that an instrument representative of those planned
for production conforms to the documented requirements
that are applicable to that instrument ([VIML] 2.5).
Examples are the →EC type examination or the →National
Type Evaluation Program.

type label
→data plate

type of protection

→explosion protection

type-specific parameters

→Legally relevant parameters with values that depend on the type of the weighing instrument. The parameters are part of the →legally relevant software and are defined during the →type approval of the weighing instrument. Examples: Parameters for the calculation of →weighing results, →stand-still lock, price calculation, →rounding of measurement results, →software identification. ([OIML R 76-1] 2.8.3)

types of approval

1. →General approval for national verification or CE verification. Measuring instrument types are given general approval for →verification if they meet the requirements of the →Verification Ordinance and recognized →engineering standards and no →type approval is prescribed in the Verification Ordinance.
2. Intrastate →type approval and →EC type approval. The national type approval/EC type approval is the approval of measuring instruments of a manufacturer after testing by the →Notified Body.

ug

Used for the unit symbol of the unit of mass →microgram
instead of →µg when Greek letters are not available (not
permissible in applications subject to legal metrology).

ultramicro balance

→Analytical balance for ultramicro analyses, with a
→weighing capacity of typically several grams and a
→readability of 0.1 µg (Fig. 172).

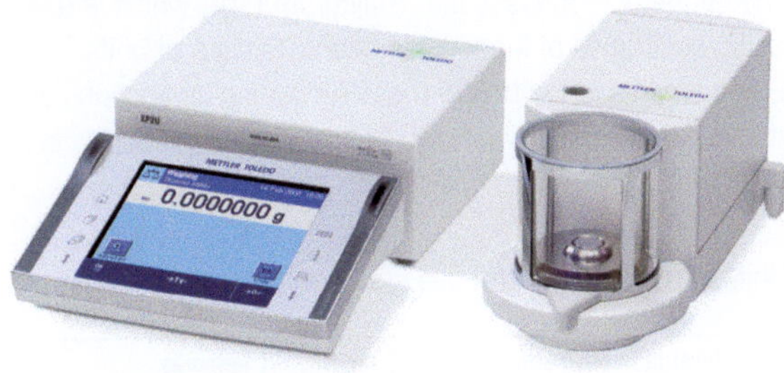

Fig. 172
Ultramicro balance
(weighing capacity 2 g;
readability 0.1 µg)
with load cell (right) and control
unit (left)

uncertainty (of a measurement)

1. Non-negative parameter characterizing the dispersion
   of the quantity values being attributed to a measurand,
   based on the information used. ([VIM:2008] 2.26)
2. Concept that describes the fact that no measurement can
   be perfect, but is always distorted by →random errors
   and unknown →systematic errors.
3. Short form for →uncertainty interval (→coverage interval). →measurement uncertainty

uncertainty interval

The →standard uncertainty u, multiplied by the →expansion
factor k

$$U = k \cdot u$$

→coverage interval

underload indicator

→overload indicator

unit

Real scalar quantity, defined and adopted by convention,
with which any other quantity of the same kind can be compared to express the ratio of the two quantities as a number
([VIM:2008] 1.9). In weighing, the most important units
are those of mass (→kilogram) and the related unit of force
(→newton). →SI units

unit conversion factor

Factor used to convert values from one unit system to another, for example: 1 kg ≈ 2.205 lb. →nonmetric mass unit

unit of force

The derived SI unit of force is the newton (unit symbol "N").
$1\ N = 1\ kg \cdot m/s^2$ →units

unit of mass

The SI unit of →mass is the →kilogram (kg), which is one of the base units of the →International System of Units. Embodiments of the unit of mass (including its fractions and multiples) are usually referred to as →mass standards, in legal metrology [18] as →weight pieces.

Units of mass in general use are:

Name of unit	Symbol	Relationship to base unit	legal unit [18]	Remarks
atomic mass unit	u	$= 1.66053886 \times 10^{-27}$ kg	•	a.k.a. 'Da' (dalton)
nanogram	ng	$= 10^{-12}$ kg		
microgram	µg	$= 10^{-9}$ kg		
milligram	mg	$= 10^{-6}$ kg		
carat	ct	$= 0.2$ g	•	only for gemstones
gram	g	$= 10^{-3}$ kg	•	not 'gr' or 'Gr'
kilogram	kg	base unit	•	
metric ton	t	$= 10^3$ kg	•	not 'T' or 'Tn'

Tab. 6
Units of mass

→Nonmetric units of mass are also in use.

unit of measurement

→measurement unit

unit switching

Device which at a keystroke allows the measurement result to be indicated in different units, switching for example between the →units of mass kg, g, ct, lb, oz, ozt, dwt.

unit symbol

Agreed abbreviation for the name of a unit, e.g. "g" for gram, "N" for newton or "m" for meter. →unit of mass

---

[18] In all countries that have signed the Meter Convention (→BIPM)

United States Pharmacopeia

1. The United States Pharmacopeia is the →pharmacopoeia
   of the United States of America (USA). The USP stipulates
   the quality standards for all prescription and over-the-
   counter drugs, food additives, and other healthcare products that are manufactured or sold in the USA.
2. Independent not-for-profit health organization with
   the same name that publishes the United States
   Pharmacopeia (www.usp.org).

units law

In Germany, short name for the Law on Units in Metrology
which defines the legal units and their abbreviations as well
as their decimal parts and multiples.

unmodifiable software

→Legally relevant software in a defined hardware and software environment which, after sealing and/or →verification,
can no longer be modified or loaded across an interface.

UPC

Abbreviation for Universal Product Code. A numbering system commonly used in the USA and Canada as a →bar
code for prepackaged food. The 12-digit UPC is part of the
→EAN code and therefore internationally unambiguous.

USP

→United States Pharmacopeia

vacuum balance

Special type of construction, usually of an analytical balance, for the purpose of performing weighments or mass comparisons in a vacuum or closed weighing chamber (Fig. 173).

validity period of verification

Nationally stipulated time period for the individual types for which the verification is valid as a function of the measurement stability of the measuring instrument and its practical use. In Germany, for instance, weighing instruments in general: 2 years; industrial scales with →*Max* > 3 metric tons: 3 years; person scales: 4 years; checkweighers: 1 year.

variability

Obsolete term for →repeatability.

variance

1. Sum of the squares of the deviations of the individual values of a set of n measurement data $\{x_i\}$ about its mean value, divided by n-1.
2. Square of the →standard deviation.

variation coefficient

Ratio of the →standard deviation to the →mean value; relative standard deviation

$$CV_x = \frac{s_x}{\bar{x}} \;.$$

vehicle on-board weighing system

→Non-automatic weighing instrument for on-board →weighing of the load of commercial vehicles, e.g. refuse trucks (→garbage scale). For this purpose, three, four, or more →load cells are built in between the chassis and the vehicle body (Fig. 174a), usually on an auxiliary frame. The influence of vehicle tilt on the weight is compensated with the aid of an →inclination sensor (Fig. 174b).

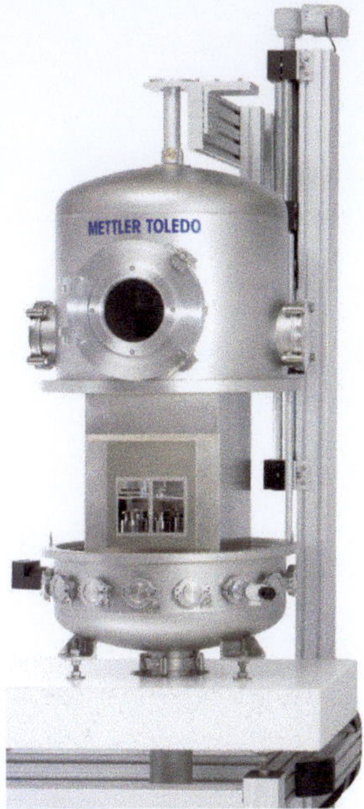

Fig. 173
Mass comparison (weighing capacity 1 kg; readability 0.1 µg) with vacuum container. Weighments can be performed at constant pressure or in a vacuum.

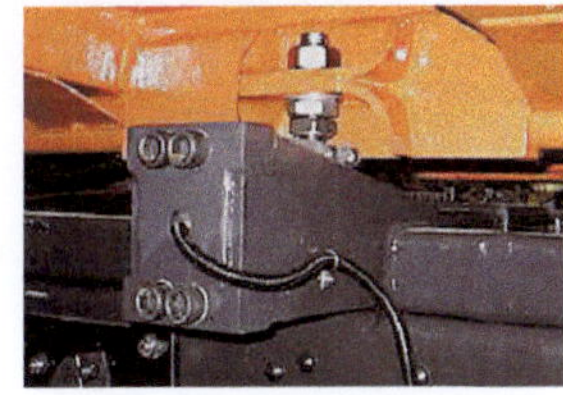

a)

Fig. 174
Vehicle on-board weighing system
a) Load cell of a vehicle on-board weighing system;
b) test of the influence of vehicle tilt on the weighing

(Image 171a by courtesy of Digisens AG, Murten, CH;
Image 171b by courtesy of GIP GmbH, Waagen- und Maschinenbau KG, Wilnsdorf, DE)

b)

vehicle scale

A →scale whose →platform is designed to weigh vehicles,
e.g. →rail scale, →road vehicle scale.

verifiable

Possessing the →ability of being verified.

verification

Procedure (other than type approval) which includes the examination and marking (→verification mark) and/or issuing of a →verification certificate, that ascertains and confirms that the measuring instrument complies with the statutory requirements ([VIML] 2.13). →EC verification

verification certificate

Document certifying that the →verification of the measuring instrument was carried out with a satisfactory result ([VIML] 3.3).

verification instructions

Official instructions addressed to →Weights and Measures authorities specifying how calibrations are to be carried out. For example verification instructions for →non-automatic weighing instruments.

verification mark

1. Mark applied to a measuring instrument certifying that the verification of the measuring instrument was carried out with satisfactory results ([VIML] 3.7).
2. Term referring to →verification with the mark for EC verification (→EC verification marking) in the EEA and Switzerland. In Germany, national verifications are identified with the →main verification mark. ([OIML R 76-1] 7.2) →stamping mark

Verification Ordinance

Decree that applies in Germany, promulgated by the authorities within the →Weights and Measures Act. The general part lists the general regulations for the approval and certification of all measuring instruments requiring verification. In annexes to the Verification Ordinance [VO], the special regulations for the type approval, definitions, requirements, inscriptions, and error limits for the individual measuring instrument types are specified. In addition, the regulation contains legally binding references to the EC directives issued for the individual measuring instrument types, as well as information and recognized engineering standards such as PTB requirements

and/or standards which include constructional and metrological requirements for the measuring instrument type.

verification procedure for weighing instruments
When verifying (→verification) weighing instruments at the
→place of installation, the following procedures may be
used:
1. Testing with full →standard load, used primarily with
    weighing instruments that have relatively low →maxi-
    mum capacities (less than 3000 kg), but also with
    →rail scales and sometimes also with →road vehicle
    scales (with →standard test vehicles) as well as
    weighing instruments that have a large →number of
    scale intervals.
2. Testing with a partly unknown load (substitute load).
    When testing with a partly unknown →load – a test
    that is used mostly for weighing instruments that have
    relatively high →maximum capacities – the →standard
    load need not be more than 1/2 the maximum capac-
    ity, or 1/2 the maximum capacity including the added
    →maximum tare load (maximum load). The standard
    load can be reduced to 1/5 of the maximum capacity if
    the repeatability error is sufficiently low, which must be
    established in advance.
2.1 Testing by the complete →step method
    In the complete step method, the weighing instrument
    is loaded with the →standard load (step 1), which is
    then replaced with a substitute load until the →meas-
    urement value is sufficiently close to the measurement
    obtained with the standard load. The standard load is
    then added to the replacement load (step 2), and this
    standard load replaced by further replacement loads
    until a measurement value is obtained that is as identi-
    cal as possible, and so on until the required maximum
    capacity has been added. This method is used particu-
    larly when the →weighing-out device must be tested for
    intermediate loads, as for example on weighing instru-
    ments with a →deflection weighing device, in that the
    standard load of a step is added gradually.
2.2 Testing by the abbreviated step method
    In the abbreviated step method, first the →standard
    load is placed on the weighing instrument, and then the
    weighing instrument is brought into equilibrium with a
    replacement load approximately equal to the maximum
    required load reduced by the standard load, and then
    the standard load is added again. The method is only
    authorized for national verification and requires the

→weighing-out device to be pre-tested, i.e. the division must be correct within specified error limits and the error of the normal ranges must be known. The abbreviated step method may only be used for →sliding weight balances and →dial weight balances.

verification scale interval

A scale interval e expressed in units of mass. When the weighing instrument is classified or verified, the value is used as a basis to determine among other things the →error limits ([OIML R 76-1] T.3.2.3). The verification scale interval is usually identical to the →actual scale interval d of the weighing instrument (exceptions include, for example, all →analytical balances). →digit, →readability

verification stamp

A stamp or seal applied by the →Weights and Measures authorities to the tested measuring instrument to attest that verification has been carried out.

verification stickers

In Germany, verification stickers with the inscription "Verified until ..." have the following colors:
1. Year with final digit 0 or 5: yellow
2. Year with final digit 1 or 6: brown
3. Year with final digit 2 or 7: blue
4. Year with final digit 3 or 8: gray
5. Year with final digit 4 or 9: green
The inscriptions are black. The marks are generally circular with a diameter of 22 or 30 mm.

vibration

Movement of the →weighing table or →support at the →place of installation caused by machines or instruments in the vicinity of the place of installation, by microseismic activity of the ground, by floor sag in buildings, building sway caused by wind or by traffic carriers (road and rail traffic).

vibration damper

1. →damping, →damping systems
2. Supporting elements on →weighing tables that damp shocks and interfering vibrations at the →place of installation by means of high internal friction. →installation of weighing instruments

vibrations

→vibrations, →string

vibrospatula

An accessory used to →weigh in granular or powdery substances. It consists of a feeding channel with a handle that can be caused to vibrate (Fig. 175). A vibrospatula allows to quickly weigh accurate sample quantities.

voltage fluctuation

Deviation of the electric power supply voltage from its nominal value of, for instance, 230 V. The admissible voltage fluctuations indicated in the operating instructions (usually -15% to +10% of the nominal voltage) do not adversely affect the accuracy or service life of the instrument.

voltage selector

In electrically operated instruments, this device allows adjustment of the instrument power supply to the existing power supply voltage, if the instrument does not operate with a voltage range that covers all power supply voltages that occur worldwide.

volume

Space enclosed by a body. The SI unit for volume is the cubic meter (unit symbol "m^3"), and the usual symbol V.

volume comparator

Device for accurate determination of the →volume or →density of a solid body (→density determination), consisting of a →mass comparator and a container that is filled with a liquid (Fig. 176). Comparative weighments of the test object with a reference object whose mass and volume are known allow the volume of the test object to be determined. Although the density of the liquid need not be known, a higher density of liquid improves the measurement effect. Determination of the volume or density by means of the volume comparator provides the most accurate results.

Fig. 175
Vibrospatula
(Image by courtesy of NeoLab
Laborbedarf-Vertriebs GmbH,
Heidelberg, DE)

Fig. 176a
Volume comparator: overall view

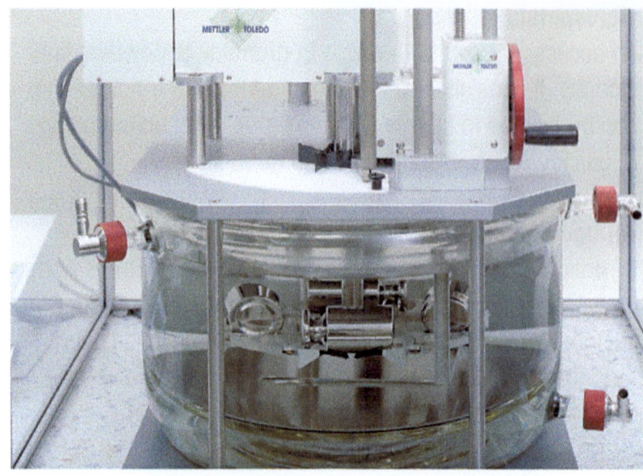

Fig. 176b
Volume comparator: liquid container

volume determination

Determination of the volume of a solid body is very similar
to →density determination.

volumetric

→volumetry

volumetric flask

Container for measuring liquids that comprises a glass flask
with a narrow neck (Fig. 177). A volumetric flask has no
scale, only a single mark to measure a nominal volume
(usually accurate to within 0.2% or less). It is used, for
example, to prepare a solution with a precisely defined volume concentration (→volumetry). Guidelines for volumetric
flasks are contained in [OIML R 4].

volumetry

Quantitative analytical method in which the volume, or a
characteristic that depends on it, is determined by measurement of the →volume; volumetric determination. →titration,
→volumetric flask, →measurement cylinder, →burette,
→pipette, →pycnometer (compare: →gravimetry)

Fig. 177
Volumetric flask
(Image by courtesy of DURAN
Produktions GmbH & Co. KG,
Mainz, DE)

warm-on time

→warm-up time

warm-up time

Time between the moment power is applied to the weighing instrument and the moment at which the instrument is capable of complying with the requirements, e.g. the →specifications guaranteed in the data sheet or, for verifiable models, within the →error limits ([OIML R 76-1] T.4.5). At the end of the warm-up time, the instrument has attained its operating temperature and thus its thermal equilibrium. For a →low-resolution weighing instrument, this usually takes less than 30 minutes; for a high-resolution weighing instrument it may take several hours. →drift

warning limit

One- or two-sided →tolerance of a process relative to its target value. Violation of this limit is not in itself an infringement of the quality requirements, but indicates drift of the process and therefore requires more intensive monitoring of the process. →control limit

water density

The density of water is approximately 1000 kg/m^3. Water is frequently used as reference liquid in hydrostatic weighments to determine the density of bodies (→density determination). The density ρ of degassed water at normal pressure (1013.25 hPa) between 0 °C and 40 °C can be determined from the following equation

$$\rho = a_5 \left[1 - \frac{(t+a_1)^2(t+a_2)}{a_3(t+a_4)} \right]$$ [19]

t water temperature [°C]

$a_1 = -3.983035$ °C

$a_2 = 301.797$ °C

$a_3 = 522528.9$ °C^2

$a_4 = 69.34881$ °C

$a_5 = 999.97495$ kg/m^3

To take account of the dependency of the density of water on the pressure (→air pressure), the density of water at normal pressure must be corrected by the following factor

$$\left[1+(k_0+k_1t+k_2t^2)(p-p_0)\right]$$ [19]

where

---

[19] [Tan]

p pressure [hPa]
$p_0 = 1013.25$ hPa
$k_0 = 50.74 \times 10^{-9}$ 1/hPa
$k_1 = -0.326 \times 10^{-9}$ 1/(hPa·°C)
$k_2 = 0.00416 \times 10^{-9}$ 1/(hPa·°C^2)

weigh in, to

Generic term for the activity of adding mass onto the →load receptor of a →weighing instrument. →weighed-in quantity

weigh module

1. That part of a weighing instrument that contains all the mechanical and electronic devices (e.g. →load receptor, force conductor, →load cell, →analog signal and data processing device, →interface for output of the →weighing result) to determine the weighing result, but which has no →display.
2. A mechanical assembly containing a →load cell used to facilitate the integration of load cells into one-of-a-kind →scales such as →tank, →hopper or →belt-conveyor scales (Fig. 178). It is designed to allow the scale structure to expand and contract, thus protecting the load cell from extraneous forces, to introduce the load correctly along the →axis of action and to safely restrain the scale. A weigh module typically consists of a base plate to interface to the foundation while supporting the load cell, a top plate as the interface to the scale structure and a suspension mechanism between the top plate and load cell. Other components provide horizontal checking and lift-off protection of the scale structure. Typically three or more weigh modules are needed to support a scale structure.

weigh out, to

Generic term for the activity of removing mass from the →load receptor of a →weighing instrument. →weighed-out quantity

weigh, to

To determine the →mass or →conventional mass (→weight) of a →weighed object. A weighment can be performed:
1.1 Statically: There is no relative movement between the weighed object and the →load receptor while weighing. Static weighing is always discontinuous.
1.2 Dynamically: There is relative movement between the weighed object and the load receptor during the weighment process.

Fig. 178
Weigh module with capacities from
7.5 t to 22.5 t

1: base plate
2: load cell
3: top plate
4: suspension and horizontal checking
5: lift-off protection

2.1 Continuously: The mass of an uninterrupted flow of
material to be weighed is determined without it being
systematically subdivided (e.g. →belt weigher).

2.2 Discontinuously: For each individual weighment, a self-
contained partial quantity is separated from the total
quantity and weighed (e.g. →hump scale).

weighbridge

1. Usually a medium- to large-sized →load receptor that
is supported from below (→top-loading) by means of
several (minimum four) supporting elements in such
manner that placing the load on the load receptor is not
hindered by suspension devices above the load receptor
(Fig. 179). Is used, for example, for →bridge scales or
→vehicle scales. →Béranger scale, →platform
2. Common term in Europe for →'vehicle scale'.

Fig. 179
Weighbridge

weighed object

General term for the item being weighed or weighed out.
→load, →net weight

weighed-in quantity

The mass of a sample (substance, reagent, etc.) determined
by weighing before it is processed (analysis, reaction,
thermal processing, etc.), usually by loading the substance
onto the weighing instrument. →net weight (compare:
→weighed-out quantity)

weighed-out quantity

The mass of a substance (sample, reagent, etc.) determined
by weighing before it is processed (analysis, reaction, thermal processing, etc.), usually by the removal of substance
from the weighing instrument. →net weight (compare:
→weighed-in quantity)

weigher

A person or operator who performs weighments as their
trade or profession. Some weighers may have professional
training in →weighing (e.g. an operator of a →weighing instrument of special accuracy). Weighers operating →public
scales (a.k.a. weighmasters) are publicly appointed and
certified (→Weights and Measures Act). They must take an
examination to demonstrate that they possess the necessary
technical knowledge.

weighing

→to weigh

weighing boat

Weighing container shaped like a small boat, and preferably made of glass, porcelain, or platinum, to accommodate weighed objects.

weighing capacity

Normally used for the →nominal value of the →weighing range, but sometimes for the →maximum capacity ([OIML R 76-1] T.3.1).

weighing card

A card made of paper or cardboard used to document printed →weighing values that have been determined by →weighing.

weighing chamber

Enclosed and protected enclosure of the →load receptor (→draft shield) in, for instance, →weighing instruments of special accuracy, also in some cases in →weighing instruments of high accuracy. Compare: →weighing room

weighing container

A →load receptor in the form of a container.

weighing deviations

→Measurement deviations that can occur during a weighment. The most important weighing deviations can be divided into four groups:

1. Changes in the mass of the →weighed object over time: Changes in the mass of the weighed object, e.g. water film (→adsorption), moisture →absorption, soiling, →evaporation, etc. Strictly speaking, these deviations are not weighing errors, since it is correct for the weighing instrument to measure the mass that is present.
2. Apparent changes in mass due to the occurrence of additional forces that act on the weighed object caused by, for example, →air buoyancy (→weighing value), convection, magnetic fields, electrostatic fields.
3. Deviations resulting from non-ideal behavior of the weighing instrument. The ideal behavior is given by the →specifications.
4. Reading error of the user (indication error).

weighing device

→weighing instrument

weighing error

→weighing deviations

weighing instrument

A measuring instrument used to determine the →mass of
a →sample (→weighed object), generally by measuring
the →force that is exerted by the sample on its support in
the gravitational field of the Earth (→weight force) ([OIML
R 76-1] T.1.1). A weighing instrument can therefore also be
used to measure force. In addition, a weighing instrument
can be used to determine other quantities that can be related
to mass or force (such as →volume, →density, content,
piece-count (→piece-counting device), →surface tension,
etc.). Weighing instruments do not indicate the mass of the
sample, but its →weighing value. At high →resolution, the
difference between the two quantities becomes visible.
→to weigh

Measuring instruments can be classified as follows:

1. Physical →measurement principle
    1.1 Direct →mass comparison, e.g. lever balance;
    1.2 →Force comparison, e.g. →electromechanical
         weighing instrument, →spring scale;
    1.3 Other →measurement principles, e.g. radiometric
         mass determination.
2. →Accuracy classes
    →weighing instrument of special accuracy, →weighing
    instrument of high accuracy, →weighing instrument of
    medium accuracy, →weighing instrument of ordinary
    accuracy
3. Type of working method
    3.1 →Automatic weighing instrument, e.g. →belt weigher;
    3.2 →Non-automatic weighing instrument, e.g.
         →microbalance.
4. Type of →display
    4.1 Weighing instrument without indicating device
         (without a scale numbered in units of mass),
         e.g. lever balance;
    4.2 Weighing instrument with indicating device,
         e.g. →counter scale.
5. Type of →equilibration of the weighing instrument
    5.1 →Non-self-indicating instrument, e.g. lever balance;
    5.2 →Semi-self-indicating instrument, e.g. dial weight
         balance with inclination range;
    5.3 →Self-indicating instrument, e.g. →electromechani-
         cal weighing instrument.
6. Type of →weighing-out device
    Examples: →Sliding weight balance, →deflection
    balance, →dial weight balance.
7. Type of →load receptor
    Examples: →bridge scale, →hopper scale.

8. Application
   Examples: →person scale, →yarn balance, scale for
   →public point of sale.
9. Terminology
   →scale, →balance

weighing instrument classes
→accuracy classes of weighing instruments

weighing instrument construction
Design of a weighing instrument determined by the technical
principle (e.g. electrodynamic compensation) or other important points (e.g. verifiable). Usually characterized by the
manufacturer as a series of types or models with a designation (e.g. name, numeric combination).

weighing instrument functions
All the functions that a weighing instrument performs. Besides the actual weighing with indication of the weighing
result, there are many other predefined or programmable
functions, such as statistical functions or →piece counting.
→operating modes of a weighing instrument, →application
module

weighing instrument of high accuracy
1. A →weighing instrument with a high resolution (→high-
   resolution) of →accuracy class Ⓘ (→accuracy classes
   of weighing instruments), a.k.a. →precision balance.
2. Strictly by definition: Weighing instrument according to
   1. that satisfies the corresponding →legal metrology
   requirements.

weighing instrument of medium accuracy
1. A →weighing instrument with a medium resolution of
   accuracy class Ⓘ (→accuracy classes of weighing
   instruments), a.k.a. commercial scale.
2. Strictly by definition: Weighing instrument according to
   1. that satisfies the corresponding →legal metrology
   requirements.

weighing instrument of ordinary accuracy
1. A →weighing instrument with a low resolution of accuracy class Ⓘ (→accuracy classes of weighing instruments).
2. Strictly by definition: Weighing instrument according to
   1. that satisfies the corresponding →legal metrology
   requirements.

weighing instrument of special accuracy

1. A →weighing instrument with a particularly high
   resolution (→high-resolution) of accuracy class ①
   (→accuracy classes of weighing instruments), a.k.a.
   →analytical balance. →macroanalytical balance,
   →semimicro balance, →microbalance, →ultramicro
   balance
2. Strictly by definition: Weighing instrument according to
   1. that satisfies the corresponding →legal metrology
   requirements.

Weighing Instruments Directive

Shortened form of →'Directive on Non-Automatic Weighing
Instruments'.

weighing method

Method by which a →weighment is performed, categorized
by:

1. The physical method, e.g. →proportional weighing
   method, →substitution weighing, or →interchange
   weighing method.
2. The type of operation, e.g. to →weigh in, to →weigh out,
   to weigh back (→back-weighing).
3. The type of process, e.g. static/dynamic or continuous/
   discontinuous →weighing.

weighing pan

→pan, →load pan, →load receptor

weighing piece

→weight piece

weighing rail insert

A section of rail that is connected to a →load cell (e.g.
→strain gage); part of a transportation weighing system.
→rail scale

weighing range

1. General: That part of the load range within which the
   weighing instrument functions correctly and indicates the
   mass of the weighed object. →weighing capacity
2. In operations subject to legal metrology requirements:
   Load range between →minimum capacity and →maxi-
   mum capacity, within which a weighing instrument may
   be used for →weighing ([OIML R 76-1] T.3.1.4).

weighing rate

→weighing speed

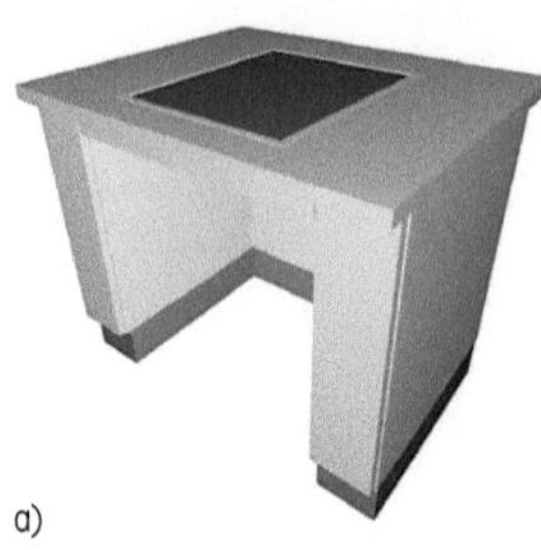

a)

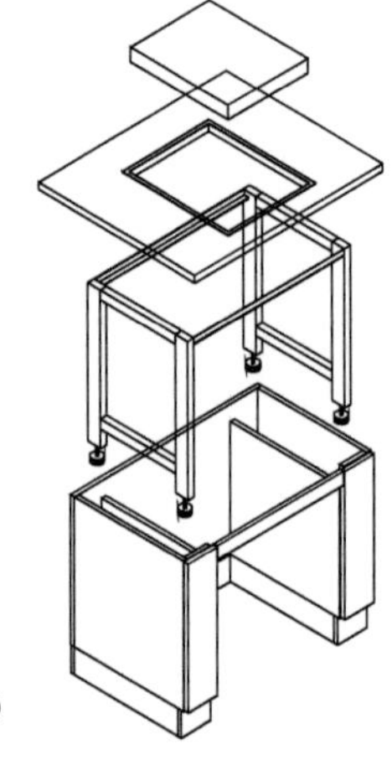

b)

Fig. 180
Weighing table
a) overall view;
b) construction

(Images by courtesy of Bense GmbH, Hardegsen, DE)

Fig. 180c
Weighing table with arm rests, specially suitable for calibrating pipettes

weighing result

The result of a weighment expressed in →units of mass or in mass as a function of time (→mass flow), mass per length (mass as a function of length, →yarn balance), mass as a function of surface area (surface load), mass as a function of volume (→density), where necessary taking into account the →weighing error and →measurement uncertainty. →Gross value, →net value and →tare value are weighing results.

weighing room

A room in which the weighing instrument is installed (→installation of weighing instruments) and weighments are performed. (compare: →weighing chamber)

weighing software

Software specially developed for weighing applications, e.g. →density determination, formula weighing, or control of inspection, measuring and test equipment. →application module

weighing speed

Number of weighments per unit of time. (compare: →weighing time)

weighing system

A →weighing instrument with a →peripheral device that combines →weighing results with other information and delivers →output signals that are used, for example, to control or adjust processes. →weighing unit

weighing table

Support on which the weighing instrument is installed, particularly when in the form of a table (Fig. 180). →Weighing instruments of special accuracy and →weighing instruments of high accuracy should be installed on special weighing tables that are as free as possible from vibrations. Monolithic stone slabs are suitable for this purpose that are either fastened to the wall or rest on two monolithic stone supports that stand on the floor of the weighing room. The →place of installation and the weighing table must be so stable that the weighing instrument indication does not change if a person leans on the table or walks into the weighing area; soft damping materials must be avoided. →installation of weighing instruments

weighing terminal

→terminal

weighing time

1. The time required for a complete →weighing, including placing the →weighed object on, and removing it from, the →load receptor. →settling time, →stand-still detector

2. →Specification: On weighing instruments with draft shields, the time from opening the →draft shield, loading the weighed object, and closing the draft shield to a sufficiently stable indication of the →weighing value (taking into account the effect of environmental conditions, configuration of the weighing instrument and weighed object). Stated in [s]. →settling time

weighing tweezers

Tweezers whose tips are made of metal or another suitable material such as hard rubber, plastic, or ceramic (Fig. 181). Used when working with →analytical balances or with →weight pieces to prevent mass changes caused by perspiration, damp hands, or heat (→weighing errors).

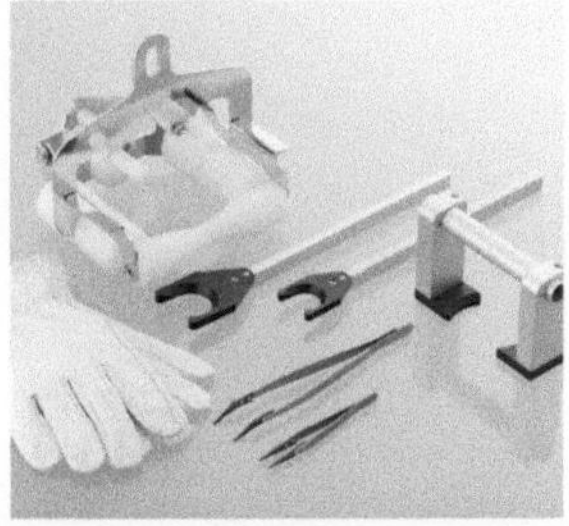

Fig. 181
Weighing tweezers and other aids for moving weighed objects and weights without touching them by hand

weighing uncertainty

→Uncertainty that limits the →accuracy of the →measurement value of a weighment. The main sources are characteristics of the weighing instrument that are described in the →specifications and are not ideal; a further contributory factor is →air buoyancy.

weighing unit

Combination of one or more →weighing instruments or →weighing systems, including the devices they control.

weighing value

1. →Measurement value that takes into account the correct zero position of the tare (zero point or tare zero) and is indicated by the weighing instrument or printed by the printer device.

2. The immediate result of a weighment before correction for →air buoyancy. The weighing value of a weighed object is equal to

   2.1 The mass of the weights that keep the weighed object in equilibrium;

   2.2 The indication W of the weighing instrument that is loaded with the weighed object of mass m

   $$W = \frac{1 - \dfrac{\rho_a}{\rho}}{1 - \dfrac{\rho_a}{\rho_c}} \, m$$

   where

m mass of the body
ρ density of the body
ρ_a density of the air ($\rightarrow$air density)
ρ_c conventional object density, 8000 kg/m^3

([DIN 1305] 3). The weighing value of air is zero.
The weighing value is different from the $\rightarrow$conventional
mass. The weighing value is not a constant quantity, since
it depends on the $\rightarrow$air density at the time of weighing [20].
If the weighment is performed at conventional density of air
(1.2 kg/m^3), the weighing value and conventional mass are
identical.

weighing-in aid
An additional coarse scale on a weighing instrument that
allows reading of the approximate mass on a moving scale
or in a special display. $\rightarrow$dispensing, $\rightarrow$available capacity
indicator, $\rightarrow$DeltaTrac

weighing-instrument-specific parameters
$\rightarrow$Legally relevant parameters with values that depend on
the individual weighing instrument. They contain adjustment
data, configuration data ($\rightarrow Max$, unit, decimal sign, etc.).
These parameters must be secured against changes ([OIML
R 76-1] 2.8.4).

weighing-out device
That part of the weighing instrument which is used in weighments to determine the $\rightarrow$mass or $\rightarrow$weighing value of
the unknown load or, when apportioning ($\rightarrow$apportion), to
set the desired mass value. On a $\rightarrow$mechanical weighing
instrument this may be a $\rightarrow$pan with $\rightarrow$weight pieces, or a
$\rightarrow$rider system.

weighment
The entirety of operations associated for carrying out, or the
objects or material involved in, a $\rightarrow$weighing operation.
([NIST HB 44] Appx. D)

weight
An ambiguous term that is used with the following
meanings:
1. As a short form for $\rightarrow$weight piece ([DIN 8120-2]);
2. As a short form for $\rightarrow$weight force;

---

[20] Besides, with $\rightarrow$electromechanical weighing instruments, it depends on the
air density at the time of $\rightarrow$sensitivity adjustment.

3. As a short form of 'mass value of a weight';
4. In non-technical language, for the result of a →weighing
   (→weighing value);
5. In non-technical language, for →mass.

Where there is a risk of misunderstanding, the respective applicable term should be used instead of 'weight' [DIN 1305].

weight class

1. Separation of →weight pieces into classes according to error limits that are specified in directives. →OIML weight classes, →ASTM weight classes, →accuracy classes of weight pieces
2. A term used in →prepackage process control to designate a specific weighing value with a defined upper and lower limit.

weight classifier

→Balance that uses the weighing result to determine a fee or price, e.g. →postal scale for postage, →garbage scale for disposal fee.

weight effect

→Weight force of a body that it exerts on its support as opposed, for example, to →spring force.

weight force

Product F_G (or often G) of the mass m of a body and →local gravity g

$$F_G = m \cdot g$$

Assuming a mean gravity →gravity) of $g = 9.81$ N/kg [21], a body with a mass of 1 kg exerts on its support a weight force of (1 kg)·(9.81 N/kg) = 9.81 N (→newton). →standard gravity

weight pan

→Pan that accommodates →weight pieces.

weight piece

Embodiment of the →unit of mass (including its fractions and multiples) that is used to determine the →mass or →conventional mass of other bodies (Fig. 182). In legal metrology, there are specific requirements for the constructional and metrological characteristics of weight pieces

---

[21] 1 N/kg = 1 m/s^2

a)

b)

Fig. 182
Weight pieces
a) OIML class E2 weights:
Right, two cylindrical weights (2 kg and 100 g), left, three wire weights (100 mg, 200 mg, and 500 mg), all from stainless steel;
b) ASTM class 6 weight:
200 g cylindrical weight, from brass, with adjusting cavity

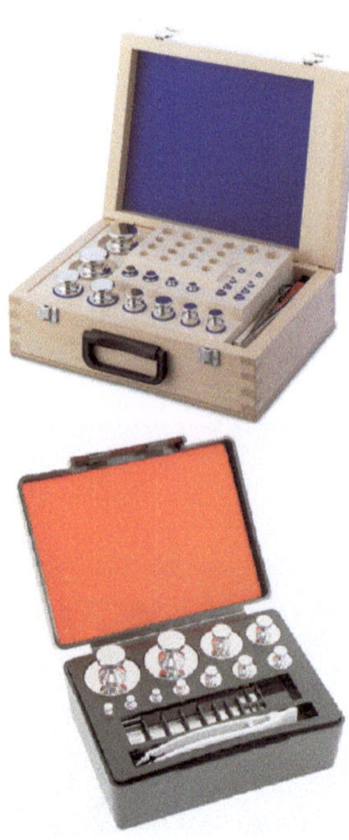

Fig. 183
Weight sets
a) OIML weight set, 1 mg to 1 kg;
b) ASTM weight set, 1 mg to 50 g

Fig. 184
Weight-dialing system
of a substitution balance

(e.g. shape, dimensions, material, surface quality, nominal value, and error limits) (→accuracy classes of weight pieces). →mass normal

weight piece, cylindrical
→cylindrical weight

weight pieces of higher accuracy class
→Directive on Above-Medium Accuracy Weights

weight pieces of medium accuracy class
→Directive on Medium Accuracy Weights

weight set
A set of several suitably assorted →weight pieces, usually decades with the denominations $(1\text{-}2\text{-}2\text{-}5)\times10^n$ (→OIML weight classes) (Fig. 183). In the United States the denominations $(1\text{-}2\text{-}3\text{-}5)\times10^n$ are also used (→ASTM weight classes).

weight unit
Term used in non-technical language for →unit of mass.

weight, specific
→specific weight

weight-dialing system
A device built into a balance and equipped with one or more →weight pieces that engage with invariable →lever arms and can be dialed from the outside by means of an adjustment system equipped with a readout (display) (Fig. 184). Weight-dialing systems may serve as the sole →weighing-out device of a balance (→dial weight balance) or may be combined with, for example, a →deflection weighing device or spring weighing device. →substitution balance

weightgrader for eggs

A device that is used to sort eggs automatically into various
→weight classes.

Weights and Measures Act

General expression for laws governing weights and measures. Such laws regulate the obligation to calibrate measuring instruments, →prepackages, volume measures,
→public scales and publicly certified →weigher, as well as
the responsibilities of the authorities.

Weights and Measures approval

→admission to verification

Weights and Measures authorities

Authorities of individual countries and states responsible for
implementing the →Weights and Measures Act and units act
including official testing of individual measuring instruments
(→weighing instruments, →weights) and supervisory tasks
(→public scales, →weigher). →prepackage process control

Weights and Measures balance

Non-technical term for a highly accurate balance used by
departments of Weights and Measures especially to test
→weight pieces of OIML classes M1 to F1 (→accuracy
classes of weight pieces) (Fig. 185).

Weights and Measures office

Term used in non-technical language for →Notified Body.

WELMEC

Abbreviation for 'European Cooperation in Legal Metrology'[22]. The cooperation relates to the metrological institutions of the member states of the European Union (EU) and
European Free Trade Area (EFTA) (www.welmec.org).

Westphal balance

→Mohr-Westphal balance

Wheatstone bridge

Electrical circuit invented by Samuel Hunter Christie (*1784,
†1865) in 1833 and improved by Sir Charles Wheatstone
(*1802, †1875) in 1843, consisting of four impedances
connected as two half-bridges. The bridge is intended for

---

[22] When the organization was founded, WELMEC was the abbreviation for
'Western European Legal Metrology Cooperation'.

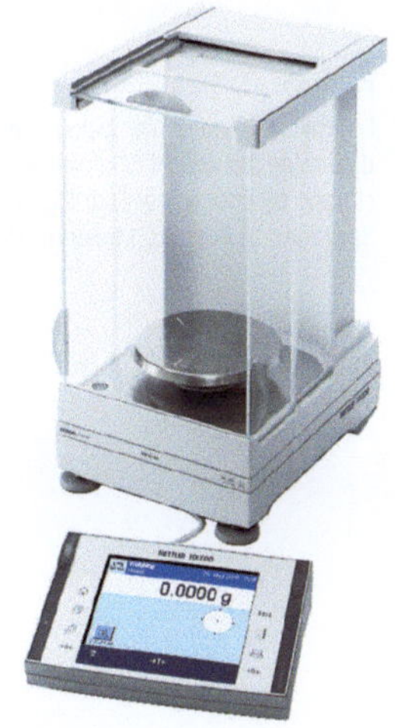

Fig. 185
Example of a Weights and Measures
balance

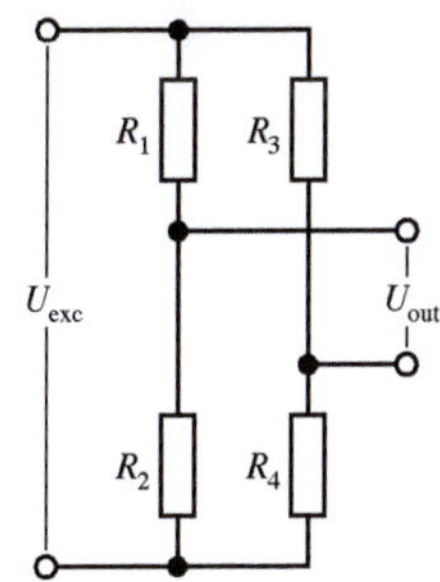

Fig. 186
Electrical measurement bridge
Four resistors, switched
as Wheatstone bridge

U_{exc}: bridge excitation voltage
U_{out}: bridge output voltage
$R_1...R_1$: bridge resistors

the measurement (→measurement bridge) of electrical impedance, e.g. resistance (Fig. 186). Due to the specific arrangement of the elements, the bridge is very sensitive. Even small changes in resistance can be measured accurately. It is therefore the preferred circuitry to measure the signal of →strain gages used in →strain gage load cells.

wheel-load scale

A →scale for determining the wheel load of vehicles. Wheel-load scales are generally used in pairs to determine the axle load (→axle-load scale) or to monitor traffic (Fig. 187).

Fig. 187
Four wheel-load scales being used to determine the weight of a vehicle (Image by courtesy of Dini Argeo, Spezzano di Fiorano, Modena, IT)

wheel-load weigher

wheel-load scale

working standard

Working standards are weight pieces that are used by, for example, verification authorities or calibration services for testing weighing instruments.

yarn balance
A balance designed to determine yarn fineness (→yarn count). →denier balance

yarn count
Measure for the quality of yarns and threads expressed as mass per unit of length. Units are the tex: 1 tex = 1 g/km (→metric unit) and the denier (1 den = 1 g/9 km). Alternatively, the quality can be expressed as length per unit of mass. Units are the metric number: 1 Nm = 1 km/kg [23] (→metric unit) and the English number:
1 Ne = 840 yards/lb.

year mark
→year mark for national verification

year mark for national verification
The year mark for national →verification in Germany consists of the last two digits of the year enclosed in a shield (Fig. 188). →main verification mark

Fig. 188
Year mark for national verification

year notation
→CE year notation, →year notation for national verification

year notation for national verification
The year notation for national verification in Germany consists of the last two digits of the year of verification; the digits are not enclosed in a shield (Fig. 189). The year notation indicates the year in which the verification was performed.
→main verification mark

Fig. 189
Year notation for national verification

---

[23] The unit symbol Nm used here is the abbreviation for "number, metric", not "newton meter"

zero indicator device

An additional →display device used to monitor the →zero position of a →weighing instrument. This device is only necessary if it is not possible to determine or set the →zero point of the weighing instrument with sufficient accuracy by using the built-in reading device of the instrument.

zero load

State of a →weighing instrument with no load on the →load receptor.

zero mark

In weighing instruments equipped with an indicator, this is the division line (scale mark, division line) that is designated "0". In weighing instruments that are not equipped with an indicator device, this refers to the opposing indicator or marking associated with the equilibrium indicator that designates the reference equilibrium position (balancing mark).

zero point

The mark (usually a line) that indicates the →zero point of the →display device.

zero point correction device

→zero-tracking device

zero point drift

→Drift of the →zero point, caused by, for example, changing →ambient temperature, by heat dissipation from the electronic components of an →electromechanical weighing instrument (→switch-on drift) or by the passage of time. →warm-up time, →drift

zero point stability

Ability of a →weighing instrument to maintain the →zero position while unloaded and to repeatedly return to the zero position even after prior loading (→repeatability at zero load).

zero position

→Equilibrium position of an unloaded weighing instrument (→zero load).

zero-setting device

Device with which the →indication of the unloaded →weighing instrument is set to zero ([OIML R 76-1] T.2.7.2).

zero-setting range

→Load range within which the →display device of the →weighing instrument is capable of being set to zero. ([OIML R 76-1] A.4.2.1)

zero-tracking device

Device that automatically maintains the zero indication within defined limits ([OIML R 76-1] T.2.7.3). Minor →zero point drifts, or material accumulation on the load receptor ($\leq 0.5\ d$), are automatically corrected when zero is indicated and the indication is in stable equilibrium, e.g. with $\leq 0.5\ d$ per second.

zone of use

An area comprising a number of →places of use for which a mean value of the →gravity can be specified for the respective weighing instrument. Depending on the →number of scale intervals of a gravity dependent instrument, it may only be used at the →place of installation, place of use, or in one or more zones of use.

Literature references [24]

[ASTM E 617]
Standard Specifications for Laboratory Weights and Precision Mass Standards
ASTM E 617-97 (2003)

[CG-18]
Guidelines on the Calibration of Non-Automatic Weighing Instruments
EURAMET/cg-18/v.02 (2009)
www.euramet.org/index.php?id=calibration-guides

[DIN 1305]
Masse, Wägewert, Kraft, Gewichtskraft, Gewicht, Last
Deutsche Norm DIN 1305 (1988)

[DIN 1319]
Grundbegriffe der Messtechnik
Deutsche Norm DIN 1319
[DIN 1319 1] Teil 1: Grundbegriffe (1995)

[DIN 8120]
Begriffe im Waagenbau
Deutsche Norm DIN 8120

[DIN 8125]
Graphische Symbole für die Wägetechnik
[DIN 8125-1] Teil 1: Grundlagen, Übersicht
Deutsche Norm DIN 8125-1 (2000)

[DIN 8129]
Selbsttätige Gleiswaagen (SGW) – Metrologische und technische Anforderungen, Prüfung
Deutsche Norm DIN 8129 (2004)
(based on OIML R 106 1)

[DIN 55350-13] Teil 13: Begriffe der Qualitätssicherung und Statistik – Begriffe zur Genauig-
keit von Ermittlungsverfahren und Ermittlungsergebnissen
Deutsche Norm DIN 55350 13 (1987)
(bases on ISO 3534)

EN 45501
Metrological aspects of non-automatic weighing instruments
European Standard EN 45501 (1992)

EN 60529
Schutzarten durch Gehäuse (IP-Code)
Europäische Norm EN 60529 (Deutsche Fassung 1991)

---

[24] Literature referenced in the text is marked by brackets '[]'.

[GUM]
Guide To The Expression Of Uncertainty In Measurement
International Organization for Standardization, Geneva (1995)
ISBN 92-67-10188-9
www.bipm.org/utils/common/documents/jcgm/JCGM_100_2008_E.pdf

[NIST HB 44]
NIST Handbook 44
National Institute for Standards and Technology, Gaithersburg, USA (2008)
ts.nist.gov/WeightsAndMeasures/Publications/H44-08.cfm

ISO 3534
Statistics — Vocabulary and symbols
Part 1: General statistical terms and terms used in probability (ISO 3534-1:2006)
Part 2: Applied statistics (ISO 3534-2:2006)
Part 3: Design of experiments (ISO 3534-3:1999)
International Organization for Standardization, Geneva

[ISO 5725]
Accuracy (trueness and precision) of Measurement Methods and Results
International Organization for Standardization, Geneva

ISO 8655
Piston-operated volumetric apparatus
[ISO 8655 6] Part 6: Gravimetric methods for the determination of measurement error (ISO 8655-6:2002)
International Organization for Standardization, Geneva

[ISO 9000]
Quality management systems — Fundamentals and vocabulary
ISO 9000:2005
International Organization for Standardization, Geneva

[IUPAC]
IUPAC Compendium of Chemical Terminology 2nd Edition (1997)
International Union of Pure and Applied Chemistry
(http://goldbook.iupac.org)

Manfred Kochsiek, Michael Gläser (Editors): Comprehensive Mass Metrology.
Wiley-VCH, Weinheim (2000)
ISBN 3-527-29614-X

[OIML R 4] [25]
International Recommendation OIML R 4
Volumetric flasks (one mark) in glass (1972)

---

[25] The catalog of OIML Publications can be looked up under 'www.oiml.org/publications'.

[OIML D 28] [25]
International Document OIML D 28
Conventional value of the result of weighing in air
(2004 (E))
(Revision of OIML R 33)

OIML R 50 [25]
International Recommendation OIML R 50
Continuous totalizing automatic weighing instruments (belt weighers)
Part 1: Metrological and technical requirements - Tests (1997)
Part 2: Test report format (1997)

OIML R 51 [25]
International Recommendation OIML R 51
Automatic catchweighing instruments.
Part 1 : Metrological and technical requirements - Tests (2006)
Part 2 : Test report format (2006)

OIML R 52 [25]
International Recommendation OIML R 52
Hexagonal weights - Metrological and technical requirements (2004)

OIML R 60 [25]
International Recommendation OIML R 60
Metrological regulation for load cells (2000)
Supplement: Certificate Transformation Requirements (2000)

OIML R 61 [25]
International Recommendation OIML R 61
Automatic gravimetric filling instruments
Part 1: Metrological and technical requirements - Tests (2004)
Part 2: Test report format (2004)
Supplement: Certificate Transformation Requirements (2004)

OIML R 76 [25]
International Recommendation OIML R 76
Non-automatic weighing instruments
[OIML R 76 1] Part 1: Metrological and technical requirements - Tests (2006)
Part 2: Test report format (2007)

OIML R 87 [25]
International Recommendation OIML R 87
Quantity of product in prepackages (2004)

---

[25] The catalog of OIML Publications can be looked up under 'www.oiml.org/publications'.

OIML R 106 [25]
International Recommendation OIML R 106
Automatic rail-weighbridges
Part 1: Metrological and technical requirements - Tests (1997)
Part 2: Test report format (1997)

OIML R 107 [25]
International Recommendation OIML R 107
Discontinuous totalizing automatic weighing instruments (totalizing hopper weighers)
Part 1: Metrological and technical requirements - Tests (1997)
Part 2: Test report format (1997)

OIML R 111 [25]
International Recommendation OIML R 111
Weights of classes E1, E2, F1, F2, M1, M1 2, M2, M2 3 and M3
[OIML R 111 1] Part 1: Metrological and technical requirements (2004)
Part 2: Test report format (2004)

OIML R 134 [25]
International Recommendation OIML R 134
Automatic instruments for weighing road vehicles in motion and axle-load measuring.
[OIML R 134 1] Part 1: Metrological and technical requirements – Tests (2006)
Part 2: Test report format (2004)

[Tan]
Tanaka M., Girard G., Davis R., Peuto A., Bignell N.: Recommended table for the density of
water between 0 °C and 40 °C based on recent experimental reports.
Metrologia 38 (2001), 301-309

[VIM:1993]
International Vocabulary Of Basic And General Terms In Metrology (VIM)
ISO, International Organization for Standardization, Geneva, 2nd edition, 1993
ISBN 92-67-01075-1

[VIM:2008]
International vocabulary of metrology — Basic and general concepts and associated terms
JCGM 200:2008 (E/F)

[VIML]
International Vocabulary of Terms in Legal Metrology (VIML) edition 2000 (E/F)
Bureau International de Métrologie Légale (BIML), Paris

[VO]
Eichordnung vom 12. August 1988 (BGBl. I, p. 1657)
(German Verification Ordinance of August 12, 1988),
as amended by Decree of February 8, 2007 (BGBl. I, p. 70)

---

[25] The catalog of OIML Publications can be looked up under 'www.oiml.org/publications'.

[Wildi]
Theodore Wildi: Metric Units and Conversion Charts.
Wiley-IEEE Press; 2 edition (1995)
www.wildi-theo.com

United States Pharmacopeia USP 30-NF 25 (2007)
(www.usp.org)
[USP<41>] General Chapter <41>: Weights and Balances
[USP<731>] General Chapter <731>: Loss on Drying
General Chapter <1058>: Analytical Instrument Qualification
General Chapter <1251>: Weighing on an Analytical Balance

Illustrations

The following companies and organizations have kindly provided one or more images for publication:

Bense GmbH	Hardegsen	DE	www.bense-laborbau.de
Bureau International de Métrologie Légale	Paris	FR	www.oiml.org/information/biml.html
Bureau International des Poids et Mesures	Sèvres	FR	www.bipm.org
Cole-Parmer Canada Inc.	Montréal	CA	www.coleparmer.ca
Digisens AG	Murten	CH	www.digisens.ch
Dini Argeo	Spezzano di Fiorano	IT	www.diniargeo.com
DURAN Produktions GmbH & Co. KG	Mainz	DE	www.duran-group.com
GAD Elektronik-Komponenten Vertriebs GmbH	Nussloch	DE	www.gad-komponenten.de
Gassner Wiege- und Messtechnik	Salzburg	AT	www.gassner-waagen.at
GIP GmbH, Waagen- und Maschinenbau KG	Wilnsdorf	DE	www.gip-waagen.de
Grimm Waagen	Tresdorf bei Korneuburg	AT	www.grimmwaagen.at
Helios & Zaschel GmbH	Mühltal	DE	www.helios-zaschel.de
Hirschmann Laborgeräte	Eberstadt	DE	www.hirschmannlab.de
Ishida Europe	Birmingham	UK	www.ishidaeurope.com
Kistler Instrumente AG	Winterthur	CH	www.kistler.com
Pfunds-Museum Kleinsassen/Rhön	Hofbieber-Kleinsassen	DE	www.pfunds-museum.de
LAUDA Dr. R. Wobser GmbH & Co. KG	Lauda-Königshofen	DE	www.lauda.de
Paul Marienfeld GmbH & Co. KG	Lauda-Königshofen	DE	www.superior.de
Vishay Micro-Measurements	Raleigh	USA	www.vishay.com/strain-gages
NeoLab Laborbedarf-Vertriebs GmbH	Heidelberg	DE	www.neolab.de
Pesola AG	Baar	CH	www.pesola.com
Pfreundt GmbH	Südlohn	DE	www.pfreundt.de
rabo - R. Bormann & Sohn	Rabenau-Lübau	DE	www.rabo-bormann.de
RHEWA August Freudewald GmbH & Co. KG	Mettmann	D	www.rhewa.com
Römermuseum Augusta Raurica	Augst	CH	www.augusta-raurica.ch
Rubotherm Präzisionsmesstechnik GmbH	Bochum	DE	www.rubotherm.de
Strack AG	Schaffhausen	CH	www.strack.ch
Trisa Elektro AG	Triengen	CH	www.trisa.ch
Ruhr-Universität Bochum	Bochum	DE	www.ruhr-uni-bochum.de/thermo
University of Washington	Seattle	US	www.npl.washington.edu/eotwash
Wikimedia Commons	—	—	www.wikipedia.org
WÖHWA Waagenbau GmbH	Pfedelbach	DE	www.woehwa.com